Technical Communication Quarterly

published four times a year by Lawrence Erlbaum Associates, Inc., for the
Association of Teachers of Technical Writing
http://www.attw.org

Correspondence concerning ATTW should be sent to Jo Allen, 128 Leazar, Campus Box 7105, North Carolina State University, Raleigh, NC 27695.

TCQ Reviewers

TCQ relies on the expertise of our reviewers not only to select articles for publication but also to help authors see possibilities for development. We thank them for their contributions to the quality of the journal.

Paul V. Anderson	Miami University, Ohio
Carol Berkenkotter	University of Minnesota
Stephen Bernhardt	University of Delaware
Ann M. Blakeslee	Eastern Michigan University
Lee Brasseur	Illinois State University
Davida Charney	University of Texas
Mary B. Coney	University of Washington
Jennie Dautermann	Miami University, Ohio
Mary Beth Debs	University of Cincinnati
Paul Dombrowski	University of Central Florida
Sam Dragga	Texas Tech University
Brenton Faber	Clarkson University
David Farkas	University of Washington
Alexander Friedlander	Drexel University
Karen Griggs	Indiana University-Purdue University (Indianapolis)
Barbara Heifferon	Clemson University
Jim Henry	George Mason University
Robert R. Johnson	Michigan Technological University
Jim Kalmbach	Illinois State University
Bill Karis	Clarkson University
Steven B. Katz	North Carolina State University
Charles Kostelnick	Iowa State University
Carol Lipson	Syracuse University
Bernadette Longo	University of Minnesota
Barbara Mirel	University of Michigan
Michael G. Moran	University of Georgia
Meg Morgan	University of North Carolina, Charlotte
Kenneth Rainey	Southern Polytechnic State University
Frances Ranney	Wayne State University
Fred Reynolds	City College of CUNY
Lilita Rodman	University of British Columbia
Mark Rollins	Ohio University
Carolyn D. Rude	Virginia Tech
Beverly Sauer	Johns Hopkins University
Gerald Savage	Illinois State University
Catherine F. Schryer	University of Waterloo
Blake Scott	University of Central Florida
Stuart A. Selber	Pennsylvania State University
Cynthia L. Selfe	Michigan Technological University
Jack Selzer	Pennsylvania State University
Brenda R. Sims	University of North Texas
Herb Smith	Southern Polytechnic State University
Rachel Spilka	University of Wisconsin-Milwaukee
Bruce Southard	East Carolina University
Sherry Southard	East Carolina University
Clay Spinuzzi	University of Texas
Dale Sullivan	North Dakota State University
Elizabeth Tebeaux	Texas A&M University
Emily A. Thrush	University of Memphis
Linda Van Buskirk	Cornell University
James P. Zappen	Rensselaer Polytechnic Institute

Technical Communication Quarterly

Volume 13, Number 3   Summer 2004

First published 2010 by Lawrence Erlbaum Associates, Inc., Publishers

Published 2010 by Routledge
605 Third Avenue, New York, NY 10017
4 Park Square, Milton Park, Abingdon, Oxon OX14 4RN

Routledge is an imprint of the Taylor & Francis Group, an informa business

ISBN 13: 978-0-8058-9537-7 (pbk)

TECHNICAL COMMUNICATION QUARTERLY, *13*(3), 245–249

Guest Editor's Introduction

James M. Dubinsky
Virginia Tech

Rhetoricians and teachers have long argued that education should prepare students for civic engagement—that is, prepare them to exercise political power by pursuing goals concerned with "human life and conduct" (Cicero I, 15) that benefit the common good (Cherry; Hauser; Sullivan). Such preparation is not without challenges, particularly in our field of technical communication. The challenges are often linked to what Dale Sullivan calls a "technological mindset" (375) or what Blyler calls a "technocratic consciousness" (142). Some challenges are connected to issues of what Jeffrey Grabill calls "involvement" (30). One of the most difficult challenges, that of balancing the view of rhetoric as a virtue linked to public deliberation and effective citizenship with one of rhetoric as a set of skills associated with job preparation, has been a topic often discussed from a variety of perspectives (Dombrowski; Dubinsky; Faber; Johnson; Sapp and Crabtree; Selber). Almost all challenges, however, involve the complex interrelation between altruism and self-interest, between a rhetorical and an instrumental approach to teaching.

One way to achieve a balance between altruism and self-interest that also illustrates both the complex nature of rhetoric and what Gerard Hauser calls rhetoric's centrality to democratic life (13) is to examine uses of rhetoric and to study it in action. In the Winter 2000 issue of *TCQ*, edited by Carolyn Rude, on the theme of "The Discourse of Public Policy," the authors did just that. The articles outlined ways that rhetoric in the form of technical communication operates within public and political spheres. That collection represents an important step in our field's understanding of the civic life of technical information and the potential of technical communicators to affect civic decisions. As Rude explains, "policies are made through discourse," and "the knowledge and interests of people in technical communication are suited to influence in policy matters" (5). The articles in that issue map roles that technical communication, and by extension technical communicators, play in civic discourse and civic life.

This special issue on civic engagement and technical communication extends that earlier work by focusing discussion more specifically on the ways educators can help students become actively engaged members of what Hauser calls a "rhetorical democracy" and defines as "a rhetorical form of governance in which all citizens are equal, everyone has a say, everyone has a vote, and all decisions are based on the most compelling arguments,...a state governed by deliberation...and the better angels of the human spirit" (1). In the first article, Cezar Ornatowski and Linn Bekins examine the concept of "community" as a locus for civic engagement and question some of the definitions of community they see embedded in current pedagogical practices, believing that those definitions are "reified, ubiquitous, always positive, and ultimately unexamined" (253). Their goal is to propose a "symbolic/rhetorical perspective on community" (253) that will shape our practice in ways that are both useful and productive.

Carolyn Rude also seeks to shape our understanding of practice. She proposes an expansion of the rhetorical canon of delivery beyond its current emphasis on publication to one that focuses on "the rhetorical situation comprehensively understood" (273). In doing so, she argues, such an extension increases the ethical demands upon writers and "involve[s] pursuit of a change in values, policy, and action" (273). Effecting that change involves "carrying the message to ever wider audiences" (282), a move that helps to promote the public good.

By emphasizing reflection and ethical deliberation in a cultural studies approach to service learning, J. Scott Blake addresses the issue of developing "students' civic awareness and engagement" while limiting the effects of "hyperpragmatism" (290). His hybrid pedagogy expands the focus from "immediate rhetorical situation[s], discrete discourse communities, and production processes to the larger cultural conditions and circulation of discourse" (299). Thus, his article addresses some of the concerns raised by Ornatowski and Bekins.

Dave Clark's goal is similar to Scott's in that he, too, is interested in "help[ing] our students gain skills and organizational awareness" and "improv[ing] the perceived relevance of our work" (311). He outlines a pedagogical model, based on a concept of engaged action research that encourages teachers and their students to work with community organizations to define their needs and create situated responses. This model addresses the concern of involvement and seeks to "promote an active vision of citizenship" and "serve the community" (309).

The final two articles approach the issue of civic engagement from slightly different angles—one examining the role of teacher as both rhetor and instructor and the other looking to the past for possible solutions for the future. Both, however, are firmly rooted in classical rhetorical theory, and both seek to address the complex tensions associated with practice and being "practical," in the sense outlined by Miller when she links "practical" with *phronesis*, a type of reasoning that she says, quoting Aristotle's *Nichomachean Ethics*, "makes one 'capable of action in the sphere of human goods'" (Miller 22). Melody Bowdon, sounding something like Cicero, who in *De Oratore* argues for a return of a system of education in which teachers give "instruction both in ethics and rhetoric" (III, XV), describes roles that teachers can play both by becoming involved in their communities as rhetoricians and by involving their students as collaborative partners. In her autoethnographic case study, she shows how technical communicators can "seiz[e] kairotic moments for intervention...by collaborating with community stakeholders and specialists in other fields" (326) and how teachers can help students become "public intellectuals" by incorporating "conscientious, ethical, collaborative models of writing instruction" (339).

Michelle Eble and Lynée Gaillet, in the final article of this issue, draw a line from Cicero to eighteenth-century moral philosophers in Scotland to point out that it is wise to look to the past in order to move into the future. By pointing out similarities between situations faced by teachers in the eighteenth century—a changing educational environment that emphasized the "increased employment opportunities... which children of working classes could not easily afford to ignore" (346)—and those faced by teachers of our own age, Eble and Gaillet highlight strategies that can help students "develop a civic mindset" (351). Both their article and Bowdon's refer to the need for students, writers, and teachers to become intellectuals who orient themselves toward the public or community good. Both articles emphasize a sense of activism that situates the work of the classroom within communities of practice.

By continuing the conversation about the relationship between technical communication and the public good and focusing specifically on pedagogical strategies and their theoretical and historical underpinnings, the authors in this special issue clarify roles that technical communication and technical communicators play in civil society, as well as ways our curricula can be shaped to prepare students to fill those roles. In so doing, they offer scholars and teachers of technical communication insight into the complex problem of education for civic engagement. We hope you will find the articles useful, and we look forward to your comments.

Jim Dubinsky

ACKNOWLEDGMENTS

As a guest editor, I have been assisted by many of the field's finest teachers and scholars. The most important help came from my co-editor, J. Harrison Carpenter, whose influence and spirit shaped our early thinking about civic engagement and ultimately this special issue. The issue began at a late night social sponsored by Michigan Tech at the 4Cs in 2000, at which I was introduced to Harrison. We began talking about our common interests in civic discourse and community engagement. Not long afterward, the idea for a special issue began to take shape when Harrison coordinated a panel on civic engagement for the next 4Cs in 2001 with Julia Williams, Pete Praetorius, and me. That panel, chaired by Carolyn Rude, led to further conversation during which Harrison and I recognized that the *kairos* was right for such an issue. Harrison helped define the topic and create an effective case with the journal's editors. Finally, he collaborated on the call for papers (and other administrative documents associated with the issue) and on narrowing the field after we received forty-three proposals in response to the call. Unfortunately, just as the manuscripts were coming in, Harrison was involved in a serious accident, which prevented him from assisting with the selection and editing of the manuscripts in the issue. Fortunately, he is recovering. I take full responsibility for any mistakes and shortcomings in this special issue.

I also want to thank the contributors whose submissions do not appear in this issue. We received many exceptionally strong manuscripts. Each one helped to further my own thinking on this issue of civic engagement, and many, I believe, will soon find their way into print. Good ideas seldom lay dormant.

Finally, I recognize the help I received from the many friends and colleagues who served as manuscript reviewers: Nora Bacon (University of Nebraska at Omaha), Steve Bernhardt (University of Delaware), Ann Blakeslee (Eastern Michigan University), Stuart Blythe (Indiana University-Purdue University Fort Wayne), Tracy Bridgeford (University of Nebraska at Omaha), Jean Bush-Bacelis (Eastern Michigan University), James Collier (Virginia Tech), Jennie Dautermann (Miami University), Paul Dombrowski (University of Central Florida), Brenton Faber (Clarkson University), Jeffrey Grabill (Michigan State University), Jack Jobst (Michigan Technological University), Susan Katz (North Carolina State University), Karla Kitalong (University of Central Florida), Melinda Kreth (Central Michigan University), Gail Lippincott (University of North Texas), Robert W. McEachern (Southern Connecticut State University), William Macgregor (Montana Tech), Carolyn Miller (North Carolina State University), Frances Ranney (Wayne State University), Louise Rehling (San Francisco State University), Barbara Roswell (Goucher College), Gerald Savage (Illinois State University), Catherine Smith (East Carolina University), Dale Sullivan (North Dakota State University), and William J. Williamson (University of Northern Iowa). I offer a special thanks to Jerry Savage, Tracy Bridgeford, and Bill Williamson, who each served as a second set of eyes at critical moments during the shortened copyediting cycle. And finally, I appreciate the forbearance both of the contributing authors, who were wonderfully responsive and gracious throughout the process, and the *TCQ* editors, Mark Zachry and Charlotte Thralls. All told, the issue, in many ways, illustrates the spirit of community that exists in our field.

WORKS CITED

Blyler, Nancy Roundy. "Habermas, Empowerment, and Professional Discourse." *TCQ* 3.2 (1994): 125–45.

Cherry, Roger. "Ethos versus Persona: Self-Representation in Written Discourse." *Written Communication* 5.3 (1988): 251–76.

Cicero, Marcus Tullius. "From *De Oratore*." Trans. E. W. Sutton and H. Rackham. *The Rhetorical Tradition*. Ed. Patricia Bizzell and Bruce Herzberg. Boston: Bedford/St. Martins, 1990. 195–250.

Dombrowski, Paul D. "Ethics and Technical Communication: The Past Quarter Century." *JTWC* 30.1 (2000): 3–29.

Dubinsky, James. M. "Service-Learning as a Path to Virtue: The Ideal Orator in Professional Communication." *Michigan Journal of Community Service Learning* 8.2 (2002): 61–74.

Faber, Brenton D. *Community Action and Organizational Change*. Carbondale, IL: Southern Illinois UP, 2002.

Grabill, Jeffrey. "Shaping Local HIV/AIDS Services Policy through Activist Research: The Problem of Client Involvement." *TCQ* 9.1 (2000): 29–50.

Hauser, Gerard. "Rhetorical Democracy and Civic Engagement." *Rhetorical Democracy*. Ed. Gerard A. Hauser and Amy Grim. Mahwah, NJ: Lawrence Erlbaum Associates, Inc., 2004. 1–14.

Huckin, Thomas. "Technical Writing and Community Service." *JBTC* 11.1 (1997): 49–59.

Johnson, Robert R. *User-Centered Technology: A Rhetorical Theory for Computers & Other Mundane Artifacts*. Albany, NY: State U of New York P, 1998.

Miller, Carolyn R. "What's Practical about Technical Communication?" *Technical Writing: Theory and Practice*. Ed. Bertie E. Fearing and W. Keats Sparrow. New York: MLA, 1989. 14–24.

Rude, Carolyn. "Guest Editor's Column." *TCQ* 9.1 (2000): 5–7.

Sapp, David Alan, and Robbin D. Crabtree. "A Laboratory in Citizenship: Service Learning in the Technical Communication Classroom." *TCQ* 11.4 (2002): 411–31.

Selber, Stuart. "Beyond Skill Building: Challenges Facing Technical Communication Teachers in the Computer Age." *TCQ* 3.4 (1994): 365–90.

Sullivan, Dale. "Political-ethical Implications of Defining Technical Communication as a Practice." *JAC* 10.2 (1990): 375–86.

What's Civic About Technical Communication? Technical Communication and the Rhetoric of "Community"

Cezar M. Ornatowski and Linn K. Bekins
San Diego State University

Although the concept of community has been advanced in technical communication as a moral reference point for civic rhetorical action, this concept is typically used in romantic, redemptive, and essentializing ways. This article argues for a radical and symbolic/rhetorical view of community, regarding it a discursive construct purposefully invoked by technical writers for strategic reasons.

"Knowledge without conscience is but the ruin of the soul."

Rabelais, *Gargantua and Pantagruel*

In the last decade, concerns with ethics in technical communication scholarship and teaching have led to broader interest in the relationship between technical communication and the quality of public life. This interest seems to have many sources. One is probably the general upsurge in interest in the problematic of discourse and democracy following the political transformations in Europe and South Africa in the early 1990s. Another is renewed concern among many educators over what Stephen Bernhardt has referred to as the growing "alienation from the polis" (600) that appears to characterize our society on the threshold of the twenty-first century. Still another is the realization that through science and technology "human beings are gaining," in the words of David Donnison, a "greater capacity to create—and destroy—their own communities, along with the values that govern their development and the environment in which they survive" (221). Attendant on this realization are concerns about social control of technology, the quality and character of technical decision-making, and the potential complicity of technical communication in the spread of what Habermas calls the "technocratic consciousness" and in what are often seen as the depredations of the increasingly global corporate economic order.

In response to such concerns, technical communication scholars have searched for ways to connect the research, teaching, and practice of technical communication to broader democratic and human concerns in order to open up the field to civic advocacy and action and make it responsive to progressive political agendas focused on individual and collective empowerment and emancipation. In a word, the goal is to "civil-ize" technical communication by disengaging it from its origins in, and bondage to, industrial-bureaucratic practice and recuperating it under the aegis of rhetoric as deliberative discursive practice, democratic in spirit and practice and focused on human needs and concerns.

Perhaps the first and most well-known attempt at such recuperation was Carolyn Miller's proposal for reconceiving technical communication as a "matter of arguing in a prudent way toward the good of the community rather than of constructing texts" (23). By *prudence*, Miller meant "the ability (and willingness) to take socially responsible action, including symbolic action" (23). The locus of such action was to be "not the individual or any particular set of private interests but the human community that is created through conduct; ... the larger community within which the corporation sells its products, pays taxes, hires employees, lobbies, issues stock, files lawsuits, and is itself held accountable to the law" (23).

Although we do not disagree with the moral intent behind Miller's proposal, we cannot help noting that the idea of community implicit in the phrasing may no longer match the reality of today's world. Large corporations such as Cisco Systems, IBM, Intel, or Qualcomm (to mention only a few familiar ones) are likely to sell products in the United States, Europe, Japan, or South America; pay taxes in the United States (or the Virgin Islands); hire employees in India, Mexico, the Philippines, or Poland; lobby in Washington and Singapore; issue stock in New York, Frankfurt, Paris, London, and Tokyo; and be held accountable to the law wherever they operate. Increasingly, many of their "communities" may also turn out to be virtual, which complicates the problem of jurisdiction, because traditional notions of legal sovereignty are connected to physical location (for a discussion of these complexities, see, for instance, Berman). This brief divagation is merely an example of the kind of reflection that made us want to explore the concept of community further. If community is indeed the locus of civic virtue that underpins rhetorical action, as it is in Aristotle's *Rhetoric*, what would be a realistic notion of community today that would best apply to, and describe, the work technical communicators do? In what sense exactly can one speak of (or in the name of) "community" in the diverse contexts in which technical communicators act today? What exactly is the foundation of the "civic" in technical communication?

These questions about community and civic engagement have not lain dormant since Miller's proposal. Many proposals followed hers, positing "community" as both the locus of virtue and the ethical reference point of rhetorical practice. The emphasis on community as the ethical horizon of discursive action is not exclusive to technical communication; it pervades the literature on service-learning (which is often referred to, as it is on our campus, as community service-learning) as well as much of the discussion of service-learning in composition. It is, as a review of this literature has convinced us, a concept that is central to many discussions of civic engagement in composition as well as technical communication. Our review has also made us aware of the degree to which the concept of "community" is typically used as a "god-term" in the sense coined by Kenneth Burke: reified, ubiquitous, always positive, and ultimately unexamined.

Our purpose in this article is to further the discussion of the civic engagements of technical communication by examining the notion of community from a variety of perspectives. We begin by examining conceptions of community offered in technical communication and composition literature, as well as service-learning. Next, we examine conceptions of community offered in sociological literature to identify a conception that may potentially be very useful to technical communication scholars and students. Finally, we review some relevant rhetorical studies of community, including some examples from practicing technical communicators, and propose a symbolic/rhetorical perspective on community that offers a more realistic and useful framework for teaching one important civic aspect of technical communication: the role discourse plays in constructing communities. We end by suggesting some implications of our proposed view of community for technical communication scholarship and teaching.

CONCEPTIONS OF COMMUNITY
IN TECHNICAL COMMUNICATION

In technical communication, community has typically been advanced as the source of civic virtue: an ethical reference point located between, on the one hand, the potentially undemocratic, impersonal, and bureaucratic state and, on the other hand, the equally impersonal technical/economic rationality of industry and the market. Such proposals, however, although following the spirit of Aristotle's *Rhetoric*, present a number of difficulties. We'll briefly discuss only a few examples.

Thomas Miller proposes to incorporate the "larger public context" into technical communication as part of an effort to base its practice on "civic humanism." The aim of civic humanism is to produce citizens with practical wisdom, or phronesis, citizens who "say the right thing at the right time to solve a public problem because they know how to put the shared beliefs and values of the community into practice" (57).

The idea of "shared beliefs and values" as characteristic of (or as the very foundation of) community is relatively common in discussions of civic education and service-learning. However, in actual social practice such an idea appears anachronistic. What would such "beliefs and values" be, for example, in a place such as our own "community of San Diego" (an expression frequently used in the local media to promote the spirit of, well, "community")? Would they be the family, Catholic, pro-immigration values of the city's sizable Latino community, with their traditional across-the-border ties to Mexico? Or perhaps the anti-growth (and sometimes anti-immigration) values of many "native," older, or poorer residents who are getting priced out of the housing market? Or the pro-growth values of the city's sizable development industry or of middle-class technocrats, to whom San Diego offers favorable business opportunities in global industries such as biotech along with phenomenal real-estate appreciation? Or maybe the patriotic, "traditional" American values of the city's sizable military establishment? Which of these values would one espouse when drafting, let's say, a proposal for the conversion of the Miramar Marine Corps Air Station (the facility made famous in the film *Top Gun*) or for the planned—and controversial—relocation of San Diego's international airport? And if one suggests that San Diego is not one community but many, how—along what lines—are they to be defined?

David Sapp and Robbin Crabtree propose that "education for citizenship" must include experience working with communities—in their case, migrant farm workers. Their rationale deserves to be examined at some length:

> There is no doubt that education should provide our students with the credentials and training to gain employment. The skills gained in traditionally taught academic courses and internships in industry and business allow students to repay loans and survive an ever-increasingly competitive world. But universities should also provide teaching, research, and service that benefit society in other ways. Service-learning projects help instructors teach students how to serve others and how to be socially responsible, and like the traditional industry internship, service-learning projects help instructors teach students how to struggle with real-world problems. (413)

Underlying this rationale seems to be a duality between the world of work on the one hand, which seems to be assumed to be inherently indifferent or hostile to community, and "serving others" through "struggle with real-world problems" on the other hand. The site of this struggle is a certain sort of community: farm workers, minorities, the poor, the homeless, and so on. Terms such as "struggle" (connotations of class struggle) suggest a romantic notion of community, in which civic-minded activists lend their skills to help the downtrodden.

James Dubinsky also sees service-learning as a pedagogical strategy through which technical communication teachers will be able to "not only teach... students practical skills but also to address the civic issues involved in using those skills" (62). Dubinsky suggests that service-learning "prepares students for the workplace in a more comprehensive way than many other pedagogical strategies because students apply what they have learned by working with real audiences" (63). In addition, students will both "learn the skills they'll need in the workplace" (63) and become "ideal orators who meet their citizenship obligations" (64). There is an assumption here of a relatively unproblematic transfer of skills between the classroom, the community, and the workplace: the community is a kind of transitional halfway house through which technical writers, after doing their time, practically automatically emerge into the workplace as civic-minded rhetoricians.

We cite these examples to identify what we see as some key assumptions shared by many such proposals. Two key moments appear to be undertheorized in most of them: the concept of community and the intellectual/pedagogical moment whereby community service presumably translates into civic and, perhaps even more importantly, rhetorical awareness. The first undertheorization seems characteristic of service-learning literature in general. The second, we suppose, may be due to the fact that the concept of community as central to discussions of civic and democratic virtues appears to have entered technical communication through the side door of service-learning, rather than through the main door of rhetorical studies, where there has always been a relatively rich, if somewhat scattered, tradition of theorizing about the relationship between discourse and community. Both undertheorizations appear in the article that did perhaps more than any other to bring service-learning to the attention of technical communication scholars.

In this article, Thomas Huckin suggests that sending students into "the community" helps students develop more civic awareness (49). The community, in this case, turns out to include "the Food Bank, local homeless shelters, various environmental organizations, the Rape Crisis Center, Easter Seals" and others. Huckin suggests that the experience "broadens the students' civic understanding" (50). However, attention is not focussed on the process by which students presumably become more aware of the civic aspects of their writing practices; such an outcome is assumed to be simply implicit in the fact that students write in "real world" contexts and that they do it in the "community." The evidence cited in support of this assumption consists mainly of the students' testimony, in which they state that they learned how to apply their writing skills in "real life" and participated in an "authentic" project. Missing, however, is the link of how, and even whether, they in fact became more ethical or civic-minded rhetoricians. One gets the impression that the students equate "real life" with "civic," an equation that is problematic. All professional communicators work in "real life" contexts, but we do not assume that they become, therefore, civically virtuous (if they did, there would be no problem to begin with). If the difference is merely one of context, do we really want to believe that the difference between a civic-minded technical communicator and one who presumably mindlessly serves antisocial or particular interests is really only that one interned in a soup kitchen or church and the other at IBM? It is useful to note that every regime, including Nazi Germany and Stalinist Russia, had extensive programs of public service for the young—for example, Goebbels's "Winterhilfe" (winter help for the poor).

Huckin (and others since) suggests that simply sending students into the community is not enough; they need guidance and reflection. Discussions with the students should focus on "the larger societal needs that these organizations are trying to address" (57). Huckin suggests that the questions such discussions should address (e.g., "Why do we have such problems in our society?") may indeed raise some social awareness, especially in students who come from privileged backgrounds. They do not seem adequate, however, to bridge the gap between that awareness and the rhetorical practices students will engage in as professional communicators. The problem may lie in the conceptualization of community that appears to be a direct import from service-learning literature, but that may not be adequate to the needs of technical writing students.

CONCEPTIONS OF COMMUNITY IN SERVICE-LEARNING

Even a cursory review of the voluminous literature on service-learning reveals many uses of "community," some contradictory, none very helpful in terms of theorizing civic engagement in rhetorical terms. In much of this literature, "community" typically refers to one of three relatively unspecific, sometimes overlapping, kinds of entities: an undefined general collectivity that functions as the positive opposite of the "ivory tower," industry, individualism, or some particular interest that a given author considers as adverse to community; something "outside" of whatever it is that is designated as not the community (here, the community may be a neighborhood, a church, or even, somewhat paradoxically, a business organization, as in some discussions of "university-community partnerships," where the community is any entity outside the university); specific economic, ethnic, religious, institutional, or interest-based groups, such as "the Latino community," "the Muslim community," "the environmental community," "the business community," and so on (Cushman; Decker and Decker; Stanton, Giles, and Cruz; Wade). "Community" may also refer not to an entity but to a relationship: a relationship of self-to-other intended to represent an alternative to self-directed individualism of capitalism, to postmodern fragmentation of the subject, or to the Foucauldian subject as a product of disciplinary practices (e.g., Rhoads).

The problems with most of these conceptions are that they are either reified ("the community") and thus do not admit political differentiation or description, tendentious (only somehow marginalized or disprivileged collectivities qualify as communities), vague (anything may be a community), or merely relational (the positive pole in an opposition).

Many service-learning discussions of community are also underpinned, explicitly or implicitly, by fall-and-redemption narratives, in which community functions as the anchor of lost civic virtues (in some accounts, these virtues are represented as archetypically American). Rahima Wade's *Community-Service-Learning: A Guide to Including Service in the Public School Curriculum* illustrates a typical grand narrative: after recalling de Tocqueville's observation that "a healthy democracy was integrally connected to a lively moral civic culture engendered by local communities and associations" (2), followed by a description of the contemporary individualistic society whose sheer size and scale of economic pursuits led to a fragmentation of "our collective identity as a people with common needs and purposes," Wade concludes that "[f]or a thriving democracy, we need a majority of concerned citizens willing to participate in decisions from the local to the national level that effect their own lives and the common good" (3). The means to foster such willingness, and thus to "resuscitate a faltering democracy" (3), is community service-learning.

Our intention here is not to argue with the general ideological assumptions behind the service-learning movement or question the validity of service-learning as an educational strategy. Indeed, service-learning does provide valuable opportunities for work and service. We do want to argue, however, that discussions of civic engagement in technical communication need to move beyond essentializing notions such as "the" community, beyond romantic visions of community as, in the words of Adrian Little, "homogeneous havens of the common good" (23), beyond nostalgia, and beyond using the concept of community simply as a secondary device in support of some other political objective, usually the critique of capitalism. Our discussions also need to be wary of the potential contradiction in, on the one hand, identifying technical communication with the depredations of technical-economic rationality (assumed to be inherently opposed to community), while, on the other hand, advocating sending students into social contexts where they are presumably to learn to resist this rationality or to adapt it to "alternative humanistic visions" (Sullivan, 379). If technical communication is inherently tainted, what's the point of taking it into contexts in which it is supposed to promote the common good? And if it is salvageable through community experiences, how exactly is it supposed to change? What exactly is the nature of the civic knowledge we expect technical communicators to have or gain?

To begin to answer such questions, technical communication scholars and teachers need to rethink the notion of community in rhetorical and discursive terms. In the remainder of this article, we want to begin this task through a three-pronged approach: by searching for conceptions of community that more realistically describe the realities of the world in which most technical communicators live and work; by consulting the rich tradition of rhetorical scholarship on the relationships between discourse and community; and by looking at some examples of how professional communicators construct communities in their work.

COMMUNITARIAN CONCEPTIONS OF COMMUNITY

According to Adrian Little, political theorist at the University of London specializing in community issues, any conceptualization of community today "must grapple with...the context of diversity and value pluralism" (7). "[W]e cannot just assume," Little argues, "that communities exist in the requisite form or that communities only need to be reactivated to perform some kind of romanticized vision of their role in previous times" (8). The "changing context of the modern world," he suggests, "undermines much of what has been traditionally thought of as the realm of community, and yet it also provides us with opportunities of rethinking the ways in which Aristotelian virtues and the common good can be realized in contemporary circumstances" (13). Such a rethinking is all the more necessary because although traditional place- or culture-bound conceptions of community no longer accurately describe our sociopolitical reality, continuing discussion and debate concerning public interest and the locus of civic-minded discursive action remains one of the best ways to engender such interest and assure its survival in the face of particularistic agendas and strictly instrumental rationalities. At the same time, "the advocacy of a multiplicity of communities and the recognition of value pluralism implies that these processes may also generate disagreement and conflict." Hence, Little emphasizes "the need to avoid eulogizing community as a source of consensual governance" (13).

Little distinguishes two major orientations in modern views of community that between them define a range of possible intermediate conceptions: orthodox communitarianism and radical views. Orthodox views of community rest on three major assumptions: that a community represents "a substructure of common values and beliefs in which social differences are overridden," that consensus can therefore emanate within a community from a "human-level" politics beyond the reach of special interests and professional politicians, and that focus on community can reach beyond specific and particular political commitments and positions, beyond, in the words of the title of a book by Anthony Giddens, "left and right." (Little 155). Orthodox views of community treat community as, in general, representative of the social, and therefore, according to Little, cannot adequately deal with the extent of diversity in contemporary lived experience of people without resorting to a "fabrication" of some underlying overall morality. The proposals for service-learning in technical communication discussed at the beginning of this article seem to rest on orthodox assumptions about community.

Radical views of community, on the other hand, attempt to offer a more realistic account of the complex composition of the modern social body. In contrast to orthodox views, they "do not assume a homogeneity in the nature of community or any kind of dominant moral voice" because "they recognize that the existence of a multiplicity of communities and associations prevents such a dominant morality from developing" (155). They also do not equate community with society, recognizing that "society comprises many different communities (as well as a range of other forms of association) that confer differing identities on the individuals therein" (155). Instead of consensus, which they regard as apolitical and "divorced from the lived experience of Western societies today, and increasingly, societies across the world" (155), radicals see the sphere of community as one of "contestation and conflict as much as it is one of agreement" (154). It is thus a sphere that is "deeply political" (154). In fact, many radical communitarians, such as Chantal Mouffe, see a central role for discourse in community politics. In this formulation, radical communitarianism appears to us not only more realistic than orthodox views as a description of contemporary social reality, but also potentially more friendly to rhetorical perspectives on community.

RHETORICAL APPROACHES TO COMMUNITY

So far, we have suggested a rethinking of the social world (including different sorts of communities) in terms of various, and to varying degrees interdependent, domains of action (economic, political, instrumental), each implying also structures and modes of sociality, rather than in terms of (physical, geographic, psychological, or other) entities. The next step is to reconceive those actions, structures, and modes in discursive and rhetorical terms.

In his well-known book, Benedict Anderson suggests that "all communities larger than primordial villages of face-to-face contact (and perhaps even these) are imagined" (6). Anderson sees "community" as a "deep, horizontal comradeship," a comradeship that is itself "imaginary," a product of symbolic action (7). A "symbolic" view of community that seems most useful from a rhetorical standpoint has been proposed by Anthony Cohen.

In his analysis of the concept of community as one of the "key ideas" of the modern social sciences, Cohen seeks to capture the meaning of community, a la Wittgenstein, not by trying to describe what it inherently may be as an object, but rather by examining the way community functions as a concept both for those who live "inside" it as well as those who deploy it for a variety of purposes.

Cohen suggests that community is a relational concept based on a perceived relation between putative members in contrast to nonmembers and/or other communities. Thus, a defining characteristic of community is the boundary between inside and outside. To the extent that this boundary is typically not physical but perceived (by members and/or others), communities can be viewed as "symbolic." Cohen suggests that the symbolic boundaries of community are largely constituted by people-in-interaction. He also suggests that the idea of community itself is largely symbolic, in the same sense in which all relational concepts are symbolic: they allow us to freely delimit areas of reality in relation to each other and to ascribe relative values to them (see also Bauman). In turn, these values underpin our attitudes and actions as long as the boundaries on which they depend maintain their salience in our consciousness, which largely means that they remain useful for some practical purpose. According to Cohen, the "quintessential referent of community is that its members make, or believe they make, a similar sense of things either generally or with respect to specific and significant interests, and, further, that they think that that sense may differ from one made elsewhere" (16). The "community itself and everything within it, conceptual as well as material," Cohen argues, "has a symbolic dimension, and, further,...this dimension does not exist as some kind of consensus of sentiment. Rather, it exists as something for people to 'think with'" (19). In addition, communities imply not only meanings and interpretation, but also vocabularies: "People put down their social markers symbolically, using the symbolic vocabulary which they can most comfortably assimilate to themselves, and then contributing to it creatively. They thereby *make* community" (28, emphasis added).

To make matters more complex, Cohen points out that even sharing a vocabulary is no guarantee of actual unanimity of meaning or attitude; two Catholics professing "I believe in God" may have quite different things in mind. He notes that a "community can make virtually anything grist to the symbolic mill" of collective identity, "whether it be the effects upon it of some centrally formulated government policy, or a matter of dialect, dress, drinking, or dying" (117). From the symbolic perspective, it thus becomes difficult to impute "beliefs" and "values" to communities. Rather, community-formation and change appear to be a function of strategic symbolic (including discursive) action conditioned and constrained by multiple factors. As Cohen points out, "[W]hether or not people behave within the 'community' mode or in some more specialized and limited way is less a matter of structural determinism than of [symbolic] boundary management" (28).

Many studies of rhetorical formation and constitution of communities exist (for some recent studies, see Bruner; Hogan; Rodin and Steinberg). As Martin Medhurst has suggested, "Rhetoric both operates within communities and by that very operation often forms new communities," and that "community formation is a continuous process" (163). Gerard Hauser describes the public sphere as reticulate (network-like), consisting of different arenas of concern, discussion, and debate, with the arenas permeable, overlapping, and connecting in ever-changing ways and patterns. These arenas are constituted by communities and publics that come into being or are brought into being around specific issues. A collectivity that may represent a "community" in relation to one issue or dimension (i.e., geographic proximity) may be rife with divisions, even violence, in relation to some other dimension (i.e., religion), or some of its elements may form alliances, in effect also communities, with elements of other collectivities. Rolf Norgaard notes, "both civility and community take on different forms for different issues and for different publics" (249).

RHETORICAL CONSTRUCTION OF COMMUNITY IN TECHNICAL COMMUNICATION

Technical writers also engage in symbolic/rhetorical construction or reconstruction of communities, both as part of their professional rhetorical activities and through other venues. Consider an example from a medical writer's workplace: a document with which most of us are familiar, the package insert that comes with pharmaceutical products. Such inserts, usually small, densely printed and multiple folded sheets, provide information for both the prescriber and the patient for the safe and effective use of the drug, as well as a basis for the uses (and limitations) for a drug's promotion to medical industry communities and the general public. This document addresses four distinct primary audiences: the physician, the pharmacist, the patient, and the regulatory agency. The insert is a summary expression of all the experimental data in the New Drug Application (NDA) collected up to the time of submission of the application to the Food and Drug Administration (FDA).

However, the construction of the insert and of the claims it makes about the drug often begins before any research has been performed—in fact, before the drug actually exists. The initial claims may not in effect describe what the drug does or how it works; they are claims that the company would like to be able to make, and they in turn drive the research and development effort. The construction of the insert is shared among the marketing staff, who construct the claims in light of the projected interests and values of the drug's (also projected) publics; the medical staff, who eventually supply the research that bears out the claims; and the regulatory affairs and legal staff, who ensure that all government requirements are met. The pharmaceutical company's medical writer who writes the insert and the voluminous documents required by the process of FDA approval must meet the expectations and respond to the concerns of various "communities" she constructs through the claims she makes: the medical community represented by users, such as physicians and pharmacists; the potential patients; the insurers; as well as the "public" as represented by the FDA. Looking over her shoulder are the company's shareholders and employees. The writer is, in fact, a member of at least two of these communities, in each of which she has a potentially vital stake, although she may contrive to manipulate the boundaries of these communities in ways that lessen or obscure her burden of choice. In summary, the package insert is initially constructed to discursively project the interests of imagined stakeholders; then it is revised continually as the research effort begins to direct which (if any) of the initial projections can be developed or supported. The final draft of the insert attempts to mediate several "community" interests, which remain in continual tension.

The matter gets further complicated once the package insert receives FDA approval and is listed on various drug directories used by insurers for coverage decisions. Consider a writer whose job it is to write guidelines for a primary audience of medical directors at an insurance company to assist them in their coverage decisions. The writer is asked to research, document, and sometimes even make insurance coverage recommendations about drug indications that have received FDA approval, as well as any off-label listings in the drug directories used for authorizing coverage under Medicaid and other insurers. The guidelines include an overview of the drug and its indications, a section with clinical study summaries to demonstrate efficacy and safety, and recommendations, including patient selection criteria. It is important to note that drug directories play a key role in supporting brand-name drug use by listing off-label uses as well as FDA-approved indications for insurance reimbursement, and that some directories include more off-label indications than others. Thus, besides the primary audience of a medical director, the writer must also consider a secondary audience of physicians and pharmacists, as well as the FDA, other insurers, various drug directories, and perhaps even the pharmaceutical company whose drug is being documented. According to a recent report in *The Wall Street Journal* about drug directories (10/23/03), if a prescribed drug is listed in a drug directory used by an individual's insurance, it is likely to be reimbursed, regardless of whether it is an off-label or an FDA-approved drug indication. This simple fact gives rise to multiple considerations about what should be listed in the directories. On the one hand, the American Medical Association supports insurance coverage for *any* prescription representing "safe and effective therapy," as long as physicians pay attention to the scientific evidence and medical opinion. On the other hand, the rise in off-label prescriptions for uses not approved by the FDA raises insurers' costs and potentially reaps significant financial benefit for the pharmaceutical industry. Such a scenario calls to question what should be included as scientific evidence and medical opinion for off-label indications and represents a potential Pandora's box. What constitutes scientific evidence? Single-patient observations? Open or uncontrolled studies? How much of this information should make it into the drug directory? How, finally, does a writer mediate between the stakes of multiple communities to address the general public's need for "safe and effective therapy"? The medical writer's problem illustrates the pluralistic contexts in which the writer constructs communities (albeit dynamic and changing ones) as part of a documentation process.

As another example, consider the work of writers for the U.S. Geological Survey (USGS). Created by an Act of Congress in 1879, the USGS stands as the sole science agency for the Department of the Interior, with a mission to serve "the Nation by providing reliable scientific information to describe and understand the Earth; [to] minimize loss of life and property from natural disasters; [to] manage water, biological, energy, and mineral resources; and [to] enhance and protect our quality of life" (USGS). The USGS functions as a public organization with a mandate to represent multiple communities in terms of its underlying goals and values and in terms of how it conducts and documents its research. Hence, a major problem for USGS writers is how to mediate between the different communities and interests that use their research and yet remain "impartial" in conveying scientific research without making explicit recommendations or competing with the private sector.

To deal with this problem, the USGS has created a "Communications Framework," which classifies its potential audiences into four "communities" and which details types of presentations of knowledge (e.g., fact sheets, flyers, journal articles) appropriate to each community and purpose. These communities are constructed based on their familiarity with science: core professionals, noncore professionals, the general public, and USGS employees. The core professionals include scientists in and outside of the USGS who are subject matter experts in the areas discussed. Noncore professionals include managers and scientists who are not necessarily earth or biological scientists, as well as knowledgeable nonscientists. The general public is composed of academics, students, media, and any other people who might have an interest in nature. And lastly, USGS employees come from a variety of disciplines ranging from biology to hydrology to geology; most have graduate degrees in their area(s) of specialization. As a direct result of these classifications, USGS documents are constructed to appeal to the projected values, purposes, and expectations of each of its audience "communities." These communities are deliberately abstract and separated from any particular exigency, locale, or issue. Moreover, USGS writers are instructed to construct "impartial" text, show no bias toward particular groups, articulate no specific recommendations or suggestions, and include no critique of the work or policies of other agencies or individuals. The USGS claims impartiality in its research and dissemination of knowledge as a way of positioning itself within the Federal-State Cooperative Program and the Other Federal Agencies Program as a scientific organization that can "provide the understanding and scientific information needed to recognize and mitigate adverse impacts and to sustain the health of the Nation's environment" (USGS, "Science, Society").

In the face of particularistic agendas and instrumental rationalities, the USGS's attempt to communicate "impartial science" is an impossibility, especially if one accepts the assumption that all communication is rhetorical and constructed according to (and to appeal to) the values, purposes, and expectations of particular audiences. The very activity of science, as well as the resulting documentation, reflects specific sociopolitical contexts. In the context of the USGS, for instance, a scientist working on gaining knowledge about suspended-sediment transport in the San Francisco Bay area may be asked to provide evaluative written comments addressing an audience of bureaucrats, environmentalists, consultants, and/or industrialists about specific public issues, such as the scientific merit behind the proposed expansion of the San Francisco airport runway. It is at this point that the writer will inevitably find herself drawn into partisan politics and have to, willy-nilly, become a rhetorician who constructs communities (with values, beliefs, interests, and so on) as an integral part of rhetorical "representation," even if she simply states that the scientific research presented is adequate in addressing a particular problem. Merit, as Aaron Wildavsky noted long ago, is always already a rhetorical concept. In replying to the question of why public matters cannot be settled on merit, Wildavsky replied that "the question presupposes an agreement on what merit consists of when the real problem is that people do not agree" (176).

The establishment of "merit" is thus precisely a rhetorical matter, and perhaps the rhetorical crux of the matter. That is why, according to Wildavsky, we have politics and public rhetoric, and where rhetorical strategies come in; as Wildavsky put it, the "ability to devise strategies to advance the recognition of merit is immensely more helpful than cries of indignation that political craftsmanship should be necessary" (177).

Thus, although the USGS writers' avowed intent is merely to "provide information so that better decisions can be made for managing natural resources" (Schoellhammer), such a statement already implicitly assumes that public resources should be managed, that decisions about their management should be informed by scientific research, and that the USGS can provide the necessary data to inform a public decision-making process. These are community values, and they are implicit in USGS writers' discursive practices, despite their attempt to remain neutral and outside the rhetorical fray.

CONCLUSION

In *The Predicament of Culture*, James Clifford calls for alternative visions of culture to ones that see the world as "populated by endangered authenticities" (a view Clifford refers to as "pure products gone crazy") (5). Clifford suggests that scholars forego "redemptive modes of textualization," and instead envision "alternative paths through modernity" that account for the vitality, adaptability, and change in actual living cultures (12). In a similar vein, our purpose in this article has been to argue that we abandon redemptive conceptualizations of community as a foundation for civic approaches to technical communication and envision alternatives that would account for the complexity, integration, interdependence, and technologization of the world in which all of us, including technical communicators, actually live and work.

To move on, we have argued for a symbolic/rhetorical view, which regards "community" as a discursive construction whose creation or invocation is always expedient in a rhetorical sense (the myth of disinterested "public" service notwithstanding). Philippe-Joseph Salazar has argued that "[e]xpediency is the heart of deliberative politics" as well as the heart of rhetoric (11). "To expedite," he suggests, "is to act with a specific use in view. Rhetorical acts, because they engage values, activate deliberation and are conducive to action, must be carefully devised and conceived and delivered so that they reach...their maximum effect" (11). Expediency here means that rhetorical acts are purposeful and calculated for practical effect. It is important to note that this is a different take on expediency than that taken in the influential article by Steven Katz, where expediency is the antithesis of community and even humanity. Such rhetorical and expedient construction of community, including—as necessary—community "beliefs and values," occurs as a regular part of technical communication practice.

Brenton Faber provides an instructive example of community construction by an outside rhetorical "hired gun." Faber describes a case where, as an outside consultant, he constructed a "meaning" for a city cemetery in a business plan in order to defend the cemetery from a hostile corporate takeover. Prior to Faber's consulting intervention, the cemetery had never conceived of itself as a certain kind of symbolic entity; it simply performed its routine functions as a repository of bodies of the deceased. To forestall the takeover, Faber constructed the cemetery as a vital element of the "community" of the city, a part of the community's identity, and thus made it a value (a value that did not exist, as he himself admits, prior to his intervention) for the community, indispensable to the preservation of the community in the face of the presumed onslaught of a faceless outside corporation (note construction of inside/outside boundary). His business plan successfully defended the cemetery from the takeover. The fact that the "faceless" (part of constructing the cemetery as community value involved putting a "face"—a community face—on it) outside corporation would perhaps have buried people more cheaply (because of economies of scale), perhaps an important consideration to many poorer citizens, was overlooked in the consultant's enthusiasm for maintaining and defending "community values." Although it is difficult not to be sympathetic to Faber's position, it is important to note (for the sake of theoretical clarity) that, although Faber appears to regard his action as implicitly justified by his intent to defend "community values" in the face of "corporate greed," this justification came from him and represented his values, not necessarily those of the city (Faber notes that the city council was indifferent to the fate of the cemetery and would gladly have sold it).

The case described by Faber shows that helping technical communication students become more civically aware may involve not only helping them understand that they construct communities as part of their professional writing activities but also teaching them to analyze such constructions through examining, among other things, "political or social topoi, patterns of metaphors, stylistic signatures, special vocabularies, and a variety of other rhetorical conventions and tendencies" (Hogan 292). It also involves training them to be sensitive to potential problems implicit in their rhetorical activities.

We believe such instruction complicates the usual view of service-learning because sending technical writing students to serve the "community" of, for example, farm workers may actually help construct this community through the students' writing in both response and in relation to some temporary exigency (much in the manner in which ethnographers to a large extent construct the peoples that they ostensibly describe, as James Clifford and George Marcus have argued in their classic *Writing Culture*). The possibility that writers construct (or co-construct) communities raises both civic and ethical issues that should not remain unacknowledged. We are not claiming that such writing may not, in a particular case, constitute a socially beneficial act for a given community, or that it necessarily does. Rather, we argue that the civic engagements of technical communication (and of rhetoric in general) consist precisely in, first, the fact that it is *through* such acts of inscription that human communities are created, maintained, or transformed, and second, that it is the ability to reflect on the various ramifications (including the ethical) of such construction in each particular case that constitutes a crucial aspect of civic awareness in the professional activities of technical communicators. From this perspective, giving technical communication students the impression that they are merely speaking for a community that somehow organically preexists their rhetorical intervention and whose values they merely have to find (although they may discover them in the rhetorical sense of invention) and articulate may actually be a serious disservice to their ability to discern the critical rhetorical, moral, and civic dimensions in what they do as professional communicators.

All communities are constructed, in many cases in and through writing. It is not, then, necessary to leave one's campus, company, or church to begin to learn to be a civically aware communicator. The (symbolic) boundary between communities, as well as between communities and something else—like the boundary between nature and culture—cuts in complex and not always apparent or predictable ways. For instance, one can see the interweaving of such boundaries (between public and private, work and nonwork, corporate and social, as well as between different types of relationships that mark these domains) in the following comment by a medical writer.

> When I write [medical] disease backgrounders, I feel a little more like I'm doing something useful; the people who read my materials (sales people, clinical scientists) will understand the disease better, and their understanding will inform their discussion of it with customers (medical professionals), but also with family and friends who ask their opinions about health questions. I recently had an experience with my sister-in-law, who had lymphoma. I edited a disease backgrounder I had written to take out technical detail and sent it to her. Then, when she started treatment, I sent her a copy of the *New England Journal of Medicine* article about a new therapy (which she was receiving). I was able to show her, in the article, that the side effects she was having were common and did not predict a bad outcome. (I didn't write the article, but I understood it as a result of my work.) That's not general public discourse, but it's "one person at a time" communication that I believe is amplified into a general public impression when it's repeated. (Susan Hudson, e-mail to authors, 1 Sept. 03)

Such boundaries are further confused and complicated by electronic communication technologies (Rheingold; Kolko; esp. Wise). A senior technical writer who is a Web designer for a major information product company in San Diego devotes his spare time as a community network writer for his neighborhood. This writer thus constructs communities both at work and at home, using the same tools and crossing boundaries. Technical communication scholars need to research and theorize the complex relationships between rhetorical actions performed by technical communicators in their various roles and the impacts that those roles may have on the various communities.

Ultimately, however, even the symbolic/rhetorical approach to community may not answer the central question: What is the source of the virtue, of ethical values, that should guide the communicator in professional practice? Perhaps that question has no final, definitive answer. Perhaps all one can do is continue to explore the persistent questions with which even rhetorical views of community leave us, questions such as the following (to paraphrase some posed by Hogan): how can the demands of technology, the dictates of economics, or the demands of national defense be reconciled with a democratic ethos? How can the different voices in the public sphere, large and small, powerful and meek, gain a fair hearing? What role do the different voices play in defining the larger culture's response to important political and social issues and how can different groups sustain their unique identities and yet contribute meaningfully to that response? How can technical communicators exercise—without recourse to easy simplifications or "merely rhetorical" palliatives—what John Lyne has called "moral imagination" in their professional rhetorical activities? Such questions should engage scholars and students in technical communication as they continue to explore their "civic" role in the constitution of democratic society. To begin to answer such questions, we need a better understanding of how technical communicators actually experience, interpret, and articulate their membership in, allegiance to, awareness of, or impact on, communities and publics and how they use (or abuse or fail, as the case may be) discourse to negotiate the variety of often conflicting interests in which their activities may be embedded. Usable theories and pedagogies of technical communication as civic discourse can come only from such an understanding.

ACKNOWLEDGMENTS

The authors thank Jim Dubinsky, as well as three anonymous *TCQ* reviewers for their insightful comments on initial drafts of this article. We also thank Mike McGraw, senior technical writer, Qualcomm, Inc.; Rita Tomlin, senior medical writer, IDEC Pharmaceuticals; Susan Hudson, Medical Writing Associates; Ellen Espiau, research science writer, PacifiCare; and David Schoellhammer and Lawrence Smith, U.S. Geological Survey, for their comments and for generously sharing their professional expertise.

WORKS CITED

Anderson, Benedict. *Imagined Communities: Reflections on the Origin and Spread of Nationalism.* London: Verso, 1983.

268   ORNATOWSKI AND BEKINS

Bauman, Zygmunt. *Community: Seeking Safety in an Insecure World.* Cambridge, UK: Polity, 2001.

Berman, Paul Schiff. "The Internet, Community Definition, and the Social Meaning of Legal Jurisdiction." In Kolko, 49–82.

Bernhardt, Stephen A. "Teaching for Change, Vision, and Responsibility." *TC* 42.4 (1995): 600–02.

Bruner, M. Lane. *Strategies of Remembrance: The Rhetorical Dimensions of National Identity Construction.* Columbia, SC: U of South Carolina P, 2002.

Burke, Kenneth. *A Grammar of Motives.* New York: Prentice Hall, 1952.

Clifford, James. *The Predicament of Culture: Twentieth-Century Ethnography, Literature, and Art.* Cambridge, MA: Harvard UP, 1988.

Clifford, James, and George E. Marcus, eds. *Writing Culture: The Poetics and Politics of Ethnography.* Berkeley: U of California P, 1986.

Cohen, Anthony P. *The Symbolic Construction of Community: Key Ideas.* 1985. Ed. Peter Hamilton. London: Routledge, 1995.

Cushman, Ellen. "The Public Intellectual, Service-learning, and Activist Research." *CE* 61.3 (1999): 328–36.

Decker, Larry E., and Virginia A. Decker. *Home, School, and Community Partnerships.* Lanham, MD: Scarecrow Press, 2003.

Donnison, David. "Rhetoric and Reality of the New Politics for the Global Age." *Progressive Politics in the Global Age.* Ed. Henry Tam. Oxford: Polity, 2001. 221–36.

Dubinsky, James M. "Service-Learning as a Path to Virtue: The Ideal Orator in Professional Communication." *Michigan Journal of Community Service-learning* 8.2 (2002): 61–74.

Faber, Brenton D. *Community Action and Organizational Change: Image, Narrative, Identity.* Carbondale, IL: Southern Illinois UP, 2002.

Gergen, Kenneth J. "Technology, Self, and the Moral Project." *Identity and Social Change.* Ed. Joseph E. Davis. New Brunswick, NJ: Transaction Publishers, 2000. 135–54.

Habermas, Jurgen. *Toward a Rational Society. Student Protest, Science, and Politics.* Trans. Jeremy J. Shapiro. Boston: Beacon, 1970.

Hauser, Gerard A. *Vernacular Voices: The Rhetoric of Publics and Public Spheres.* Columbia, SC: U of South Carolina P, 1999.

Hogan, J. Michael, ed. *Rhetoric and Community: Studies in Unity and Fragmentation.* Columbia, SC: U of South Carolina P, 1998.

Huckin, Thomas N. "Technical Writing and Community Service." *JBTC* 11.1 (1997): 49–59.

Katz, Steven B. "The Ethic of Expediency: Classical Rhetoric, Technology, and the Holocaust." *CE* 54.3 (1992): 255–75.

Kolko, Beth E., ed. *Virtual Publics: Policy and Community in an Electronic Age.* New York: Columbia UP, 2003.

Little, Adrian. *The Politics of Community: Theory and Practice.* Edinburgh, UK: Edinburgh UP, 2002.

Lyne, John. "Rhetoric and Scientific Communities." In Hogan, 264–83.

Medhurst, Martin J. "Martial Decision Making: MacArthur, Inchon, and the Dimensions of Rhetoric." In Hogan, 145–66.

Miller, Carolyn R. "What's Practical About Technical Writing?" *Technical Writing: Theory and Practice.* Ed. Bertie E. Fearing and W. Keats Sparrow. New York: MLA, 1989. 14–32.

Miller, Thomas P. "Treating Professional Writing as Social *Praxis.*" *JAC* 11.1 (1991): 57–72.

Norgaard, Rolf. "The Rhetoric of Civility and the Fate of Argument." *Rhetoric, the Polis, and the Global Village.* Ed. C. Jan Swearingen and Dave Pruett. Mahwah, NJ: Lawrence Erlbaum Associates, Inc., 1999. 247–53.

Rheingold, Howard. *The Virtual Community: Homesteading on the Electronic Frontier.* New York: Harper Perennial, 1993.

Rhoads, Robert A. *Community-Service and Higher Learning: Explorations of the Caring Self.* Albany, NY: State U of New York P, 1997.

Rodin, Judith, and Stephen P. Steinberg, eds. *Public Discourse in America: Conversation and Community in the Twenty-First Century.* Philadelphia: U of Pennsylvania P, 2003.

Salazar, Philippe-Joseph. "Life, Inc.: A Rhetoric Lesson." Inaugural Lecture, delivered at the University of Cape Town, 8 May 2002. New Series No. 228. Cape Town, South Africa: University of Cape Town Department of Communication and Development, 2002.

Sapp, David Alan, and Robbin D. Crabtree. "A Laboratory in Citizenship: Service-learning in the Technical Communication Classroom." *TCQ* 11.4 (2002): 411–32.

Schoellhammer, David. E-mail to authors. 3 Sept. 2003.

Stanton, Timothy K., Dwight E. Giles, Jr., and Nadinne I. Cruz. *Service-Learning: A Movement's Pioneers Reflect on Its Origins, Practice, and Future.* San Francisco: Jossey-Bass, 1999.

Sullivan, Dale L. "Political-Ethical Implications of Defining Technical Communication as a Practice." *JAC* 10.2 (1990): 375–86.

U.S. Geological Survey (USGS). Department of the Interior. 2002. Home page. <http://www.usgs.gov/stratplan/vision.html>.

—. Department of the Interior. "Science, Society, Solutions: An Introduction to the USGS." USGS memo FS 010-01. 3 Nov. 03. <http//:www.usgs.gov.aboutusgs.html>.

Wade, Rahima C., ed. *Community-Service-Learning: A Guide to Including Service in the Public School Curriculum.* Albany, NY: State U of New York P, 1997.

Warner, Michael. "Publics and Counterpublics (abbreviated version)." *Quarterly Journal of Speech* 88.4 (2002): 413–25.

Wildavsky, Aaron. *Politics of the Budgetary Process.* Boston: Little, Brown, 1964.

Wise, J. Macgregor. "Community, Affect, and the Virtual: The Politics of Cyberspace." In Kolko, 112–33.

Cezar M. Ornatowski is Associate Professor of Rhetoric and Writing Studies at San Diego State University, where he directs the Master's Program in Rhetoric and Writing Studies. His current research interests include rhetoric and technology, political and civic rhetoric, and especially rhetorical aspects of political transformation, with a focus on Central/Eastern Europe and Africa.

Linn K. Bekins is an assistant professor in the Department of Rhetoric and Writing Studies at San Diego State University, where she directs the Advanced Certificate Program in Technical and Scientific Writing. Her research interests focus on technical communication practices, the rhetoric of science, collaboration studies, and adult learning. She can be reached at lbekins@mail.sdsu.edu

TECHNICAL COMMUNICATION QUARTERLY, *13*(3), 271–288

Toward an Expanded Concept of Rhetorical Delivery: The Uses of Reports in Public Policy Debates

Carolyn D. Rude
Virginia Tech

Preparing students for civic engagement requires new knowledge about the uses of documents for advocacy and social change. Substantial social change results from repeated rather than from single rhetorical acts. Reconsideration of the rhetorical canon of delivery suggests expanding the concept beyond its present connection to publication (visual design, medium) to a rhetorical situation comprehensively defined. Delivery may take place over time and embrace a web of activities including field work, updates, and interconnections with other publications.

The preparation of students in technical and professional communication for civic engagement presumes their participation in social action as citizens but also perhaps as professionals employed for their expertise in communication. Students can contribute their knowledge of rhetorical situations, audiences, genres, media, and language to the tasks that engage citizens in debate on policy. As volunteers and professionals, they can participate in the research, debate, and strategic action to promote policy changes determined to serve the public interest. Their understanding of rhetoric as skills but also as a means of reasoning about issues that shape society and as a means of persuading audiences to adopt positions may empower them for leadership as well as for participation and support roles.

Technical communication students are already prepared for support roles in civic settings as editors and grant writers, but these roles do not exhaust the potential of these students for the pursuit of effective public policy either as citizens or as professionals. Achieving that potential requires knowledge of the process of social change and understanding of how rhetoric may drive that process. Exploring their potential for civic engagement requires an examination of rhetoric in civic settings. Although discourse in corporate and academic settings has been well contextualized, the field of technical and professional communication knows less about discourse as it defines social needs and promotes social change. Classroom instruction and the corporate workplace focus on short-term projects that have stopping points and fairly predictable consequences. But social change is a long-term process that may span years. Adapting the knowledge developed in corporate and academic settings will require awareness of the long-term nature of social change and the incremental nature of rhetorical acts. The concept of rhetoric itself may expand beyond the usual classroom focus on individual instances (the document, the speech) to accommodate persuasion over time: delivering a message repeatedly and in different media, actively seeking out audiences, and promoting action in response to the message. The publication is not an end in itself but a means to an end of change in policy and behavior.

Some work in the field has revealed the long-term nature of social change and the role of reports within the process. For example, John Brockmann traces the history of legislation in the nineteenth century to solve the problem of frequent steamboat explosions. Although a series of technical reports, one described as "near perfect," detailed the problems and proposed solutions, it took 28 years to reach a satisfactory resolution. No single rhetorical act persuaded legislators to take the necessary action, and the solution was stalled at various points. More influential than a single report was the cumulative effect of multiple reports and other initiatives over time. Events beyond the texts, including continuing explosions and a changing sense of the role of government, made the legislators receptive to the information that the reports had presented. Contemporary problems are just as slow to be resolved. Kelli Cargile Cook studied a 1994 report of the General Accounting Office on sexual harassment of women in the service academies. This 1994 report followed several reports on the topic in the 1980s, but the problem has reemerged in the early twenty-first century and will no doubt be restudied before the problem is solved. These histories, and others like them, suggest that reports are necessary for framing the issues and presenting the information that persuade decision makers. They also confirm that one excellent argument well presented may be just a piece of a strategy.

An argument for looking beyond the moment of the text is offered by Danette Paul, Davida Charney, and Aimee Kendall, who study the rhetoric of science. They observe that a rhetorical analysis of a text may not reveal much about the impact of the text. The analysis may describe rhetorical strategies, but without studies of the reception of the text over time, the analyses tell little about the effectiveness of rhetorical strategies. Paul, Charney, and Kendall are interested in reception by audiences, another part of the question this article raises about what rhetors do to reach audiences, but like this article, theirs insists on looking beyond individual texts to their impact over time and their connections with other related texts.

Whether change requires time or has a short horizon, rhetoric is a means of developing information and reasoning about policy choices. Because people have competing viewpoints, rhetoric is also about persuading audiences to accept and implement change. However, the methods of rhetoric are often applied to guide the development of single instances of discourse rather than the development of change over time. For example, the canons of rhetoric (invention, arrangement, style, memory, and delivery) can describe and guide the process of researching, developing, and presenting a document or speech. What such an application emphasizes is the rhetorical situation as a relatively short moment in time. What it does not foreground is the situation in which multiple documents and other rhetorical acts may work together to change values and policies. When change is complex, the work of rhetoric—invention, reasoning, presentation, and persuasion in the interest of establishing good public policy—requires vision beyond the single document. Rhetorical theory is robust enough to accommodate a long-term process of change and not just the single instance. But first the rhetorical situation must be understood as long-term, comprehensive, and complex. Understood in this manner, the work of rhetoric is not complete when the speech is delivered or the document published. Rather, delivery may mean the beginning of new work and even the motive to produce it.

The whole process of rhetoric might be reexamined from the long-term view, but the last stage—delivery—illustrates well how the process of rhetoric continues even when a single piece of discourse may seem complete. Delivery understood as taking the document and its argument to the audience may in turn enlighten the earlier stages of development, including invention. In this article, I consider the canon of delivery as it may apply to the rhetorical situation comprehensively understood to involve pursuit of a change in values, policy, and action. Social changes may require investments of time and money, new infrastructure, and new beliefs. Thus, they are not undertaken lightly, and the challenges to rhetors are substantial. Strategies of change may not be contained in texts, even though the texts are integral to the strategies. I begin with a review of the work in the field on delivery especially as the work considers publication and ethics. I then follow some activities of the Union of Concerned Scientists (UCS) as it works to encourage policy makers to convert from coal to renewable sources of energy. Its reports are crucial to the initiative, but they are interesting for this analysis less as publications than as components of a strategy to influence an audience of state and local governments and utility companies. My method is less text analysis than analysis of a situation in which texts play critical roles. My aim is to reinvigorate the concept of delivery to release its explanatory power and in turn to add to understanding of the rhetoric of civic engagement. In suggesting an expanded concept of delivery, I do not wish to abandon the concept of delivery based on publication but rather to stretch this concept to accommodate related rhetorical acts over time.

CURRENT CONCEPTS OF DELIVERY IN TECHNICAL AND PROFESSIONAL COMMUNICATION

The canons of rhetoric have been valuable in writing instruction to explain the process of developing discourse and to validate the intellectual work of writing.

Of the five canons, the final two have been least comfortably applied to print and online discourse because of their obvious relationship to speech. According to Kathleen E. Welch ("Delivery"), Greek and Roman rhetoricians regarded delivery as significant, and Cicero called delivery "the most important aspect of a speech" (217). But as John Frederick Reynolds observes in the preface to his 1993 collection, *Rhetorical Memory and Delivery,* these "problem canons" of memory and delivery "have never received the kind of widespread critical attention they deserve" in contemporary times (vii). Still, only four of the eleven chapters in his collection focus on delivery. Had Reynolds's focus been technical communication rather than composition, he would have found even less critical attention. And, although memory can claim some contemporary book-length studies (Carruthers; Yates), delivery claims no such interest unless one counts books on document design.

In speech, delivery involves voice and gesture. Contemporary efforts to apply delivery to publication in technical communication have focused on visual design (Connors; Dragga; Dragga and Gong) and medium (Bolter; Welch, *Electric*). Delivery as visual design includes typefaces and paper choices, page layout, and use of visuals. Delivery as medium recognizes options of video, electronic communication, and communication technology as well as print. These analogies to visual design and medium reinforce the important concept that until an idea becomes public (through publication or through delivery of a speech to an audience), it cannot influence an audience to act. The presentation of content influences its availability and reception. Performance has the power to make concepts understandable and to convey urgency. Delivery is essential to persuasion.

Visual design and medium suggest a definition of delivery tied to publication, whether in print, online, or video. Publication-based definitions of delivery in written discourse tacitly imply that publication is the goal and end of the writer's work. The canons themselves, as they represent development and presentation of discourse, suggest beginning and ending points in the process of investigation, composition, and publication. What is missing from such definitions is a sense of where these rhetorical acts fit into rhetoric broadly conceived as influencing policy over time. In rhetoric, a speech (and by extension a publication) is a response to an exigence, an urgent need that discourse can resolve (Bitzer). The speech or publication is thus a means to the broader ends of deliberating and influencing policy, ensuring justice, and establishing the values of a society. What happens after the speech or the publication is the measure of its worth and success: Has the exigence been resolved? Delivery understood as a finite act, ending with the performance or publication, neglects (or at least does not emphasize) the impact of the publication on the rhetorical situation, the exigence that called the publication into being.

Whether rhetoric helps to solve problems or manipulates and deceives audiences is a centuries-old question, and delivery is a target of the question. In some articulations of the canons of rhetoric, delivery has been disparaged as "mere" presentation. According to George Kennedy, Aristotle proposed consideration of delivery as a part of rhetoric, but he was nevertheless ambivalent about delivery because of its association with acting (90). In Book 3, Part 1, of *On Rhetoric*, Aristotle writes,

> An *Art* concerned with [the delivery of oratory] has not yet been composed, since even consideration of *lexis* was late in developing, and delivery seems a vulgar matter when rightly understood. But since the whole business of rhetoric is with opinion, one should pay attention to delivery, not because it is right but because it is necessary, since true justice seeks nothing more in a speech than neither to offend nor to entertain; for to contend by means of the facts themselves is just, with the result that everything except demonstration is incidental; but nevertheless, [delivery] has great power, as has been said, because of the corruption of the audience. (218)

Aristotle regrets that facts do not speak for themselves but recognizes that delivery gives power to the facts and is necessary to persuasion. His use of *vulgar, right, justice*, and *corruption* in his description reveals the close relationship between ethics and the concept of delivery.

Contemporary writers have shared Aristotle's qualms about the possibilities of distorting fact with performance (though less about the corruption of audiences) and have asserted the ethical dimension of delivery. Robert Connors, for example, links delivery and ethos, which he equates with the image of the author. He observes that image may support or sabotage the author (66). Connors does not develop the implications of image, but image is far from the superficial representation of character or performance tricks that the term may imply. The ethos of the message bearer influences the reception of the message. Ethos may derive from standing in the community or from credentials, academic or experiential, but it also requires assurance of the independence of the speaker or writer. Conflicts of interest or incentives to compromise inquiry will damage the credibility of the message. For example, if members of the commission investigating the *Columbia* tragedy have ties to NASA and the commission reports to NASA, the commission's findings may not have the power that they would if the members are independent and report to the president. Delivery by a rhetor whose ethos is discredited compromises the message.

Also concerned about the possibility that delivery may distort, Sam Dragga explores student and professional attitudes about accuracy and integrity in representational visuals and typography. For example, to fit a résumé on one page, may the author reduce the type size and leading? Both students and professional writers think this practice is ethical. But they agree that it is unethical for a company to construct a photograph of employees showing one employee in a wheelchair even though the company employs no persons with disabilities. Dragga observes that the ethics of delivery is rarely a topic in the technical communication literature and concludes, "The canon of delivery…imposes ethical obligations on writers and on writing teachers" (94). Learning that students and professionals draw different conclusions about the ethics of various situations, Dragga proposes that teachers should introduce students to guidelines on ethical delivery and to examples that will help them develop their understanding of the ethical implications of delivery choices.

Both Connors and Dragga focus primarily on the way in which the text is accurate. Theirs is a modernist perspective. But Welch considers delivery through a postmodern lens. She regards the suppression of memory and delivery in the five canons as destructive and ideological. She asserts, "The erasure of memory and delivery in the majority of student writing textbooks constitutes the removal of student-written language from the larger public arena" ("Reconfiguring" 18). The focus on invention, arrangement, and style supports the current-traditional model of writing instruction, with its priorities on the text and correctness. Limiting the realm of rhetoric to the three canons restricts its function and risks making composition an academic exercise rather than an opportunity for civic engagement. Welch's focus on ideology suggests that accuracy of visual representation and the appropriate medium may not, by themselves, fulfill the role of delivery to make the document persuasive and therefore effective in resolving an exigence.

The current studies of delivery confirm its ethical dimension and the priority of accuracy in visual representation of information, but delivery may be more complex than its publication-based definitions imply. The current definitions raise issues of image, access to information, and responsibility for establishing a voice in the public arena as well as to visual design. The need for effective delivery may raise suspicions of corruption of truth, but the alternate perspective is that delivery may be a vehicle of justice. Memory and delivery both represent the connections of the speech or publication to the world beyond the text.

The history of activities by the UCS to change energy policy suggests that delivery should be understood not just as publication but also as a series of strategic actions in which publications are crucial. Publication is a means of outreach to an audience, but although publication alone is a passive method of outreach, the UCS activities demonstrate the rhetor taking the material directly to the audience. From this perspective, delivery is outreach after publication and interconnections of the report with other documents and activities. In the analysis that follows, I consider briefly how the report that enables outreach respects principles of delivery understood as visual design and medium. The focus of the analysis, however, is on the exigence of the need to change energy sources for the sake of health, economy, and sustainability of environmental quality and on the uses of reports in pursuit of that goal.

REPORT TO CHANGE ENERGY POLICY: THE
PUBLICATION PERSPECTIVE

The Union of Concerned Scientists describes itself as "an independent nonprofit alliance of 50,000 concerned citizens and scientists across the country," whose mission entails joining science and advocacy: "We augment rigorous scientific analysis with innovative thinking and committed citizen advocacy to build a cleaner, healthier environment and a safer world" (Website). Its five initiatives are clean vehicles, global environment, food and environment, global security, and clean energy. As part of its work, it conducts research in these areas and publishes numerous reports of findings. The report I have followed falls into the clean energy category. UCS has reported the history of this document in its newsletter (*Nucleus*), on its website, and in e-mail messages to members. Thus, UCS provides a paper trail of the report as used and interprets its influences. UCS information is the source for claims in this article about its uses.

In 1993, UCS published *Powering the Midwest: Renewable Electricity for the Economy and the Environment* (Brower). In 118 pages of text and 88 pages of appendixes, the report explores alternatives to coal and oil, including wind power, biomass, and solar power. It is technical in content with numerous graphs and maps illustrating wind patterns and biomass resources and numerous tables comparing the cost and emissions of various sources of electrical power. The appendices assess energy sources and potential for renewable resources in twelve midwestern states. The report is obviously the work of scientists with expertise in designing research and gathering data. It is intended to enable utility companies, state and local governments, and citizen groups to make decisions about electrical power that will reduce dependence on oil and coal, and the research is thorough and detailed. The research and the arrangement of findings in the report represent the first two canons of rhetoric. A previous article (Rude) provides some textual analysis of the report and hints, without details, at its uses.

The report is rhetorical as well as technical, relying not on the detached presentation of data but rather on developing good reasons for readers to pay attention to the data. These reasons relate to politics (new resources will reduce dependence on oil from the Middle East and therefore reduce the chances for armed conflict); to health (emissions from coal cost the states because of illness and lost work hours related to pollution); and to the midwestern ethos (its heavy reliance on coal, its abundant renewable resources, its position as America's "heartland"). Its primary reason, however, is economic: the report shows the increasing economy of renewable resources compared with fossil fuels, and it offers the promise of economic growth as the new technologies develop. This economic reason counters the widespread assumption that environmentally sound policies are not sound economically. The argument is tailored to the audience. Growth of the economy is not

UCS's motivation for the research and report, but in order to overcome the barrier of resistance to clean energy because of the presumed cost, UCS must hear those alternative arguments and show that the assumptions are wrong.

With the data provided in the report, states should have concrete information for planning and developing clean energy resources. The data enables the states to move beyond a concept or an exhortation: the report provides the information that enables action. The report answers questions about feasibility of conversion to renewable resources by analyzing the renewable resources in the Midwest. As incentive to states, local governments, and utility companies to consider energy data seriously, it provides economic data and pinpoints locations for wind turbines and biomass harvesting.

Although the report is not slick like a corporate annual report, it uses good principles of visual design. The cover graphic illustrates the various sources of renewable energy: horizontal white lines represent wind, diagonal yellow lines represent sunlight, and green triangles represent the forests that are sources of biomass (see Figure 1). Inside, the headings are functional and informative, and the type for the body text is a readable 11-point Times Roman. Shades of red and blue add visual appeal to the maps and graphs (although coal as an energy resource is always represented with a smoggy gray). The report is perfect bound. The visual design choices are strategic: they are functional and attractive, but publication costs have been minimized in line with stewardship both to the citizen donors and to the environment.

FIGURE 1   Cover of the 1993 UCS Report. © 2001 by Environmental Law and Policy Center. Used with permission.

The graphs and tables are composed according to good principles. The structure of tables is clear; segments in pie graphs correspond to their mathematical measures; and labels are informative. The visuals seem to meet the test of accuracy. Three chapters have an obvious rhetorical purpose that could raise questions about the objectivity of the science: Chapter 1, "Overview;" Chapter 6, "The Case for Renewable Energy;" and Chapter 7, "Policies for a Renewable Future." However, the data chapters and the appendices foreground information over argument, science over advocacy.

From a publication perspective, this report seems to fulfill the imperative of rhetoric and the principles of delivery to be accurate and just. The ethos of the UCS draws on the ethos of science, as does the data. But will publication alone change policy and practice? UCS did not think so, and the activities that accompany publication show uses of the report in the field as well as follow-up and related publications. The report is a tool of social action as well as a publication, and delivery is strategic action as well as visual design and medium.

REPORT TO CHANGE ENERGY POLICY: THE SOCIAL ACTION PERSPECTIVE

Powering the Midwest: Renewable Electricity for the Economy and the Environment offers an example of practice that shows how limited it is to think about a report only in terms of publication. This report is entwined both with field work and with other publications: it helped launch some UCS advocacy activities, and its information was republished in a variety of other reports and documents.

Field Work and Advocacy

UCS employs field workers as well as researchers and writers. (Sometimes the same person fulfills all the roles.) In Minnesota and Iowa, UCS worked in alliance with local groups, extending the influence of the report through information sharing and persuasive discourse. Michael Tennis, a coauthor of the report, was the field worker in Minnesota, where UCS linked in 1996 with some citizen groups and some individuals within two rural electric co-ops to develop a plan for offering customers the option of getting their energy from wind. The pilot project included construction of three turbines by one co-op. In addition, UCS worked with some regional citizen groups in public education. As a result of their work, 1,200 customers committed to wind energy with "green pricing." This project has developed as more of the co-ops in the region have expressed interest and are following the example of the first two co-ops. The report is both a source of information regarding development of renewable sources of energy and an argument reassuring the co-ops of the economy of their choices.

Reaching out to the report's audience took place in multiple sites. In 1995, almost concurrently with the Minnesota initiatives, a 1983 Iowa law mandating utilities to generate some of their power from renewable sources was in danger of being repealed. The repeal was considered a certainty until UCS and local citizen groups started contacting legislators. The report was useful in giving activists specific arguments and facts so that they could convince legislators to vote against repeal. The repeal effort failed by one vote. The Sustainable Energy for Economic Development (SEED) Coalition was formed as a result of this work. Preparing for the next legislative session, the SEED Coalition developed a legislative proposal to strengthen the bill and worked throughout the session to prevent efforts to weaken or repeal the renewable energy law. The Utilities Board ordered utilities to sign contracts within a specified time (Levison).

What difference did the report make to these activities and their outcome? Michael Tennis, the UCS energy analyst who coauthored the report and who was the main coordinator of activities in the Minnesota electric co-op project, said this: "This outcome—a real-life project—is why we worked so hard on *Powering the Midwest*, and it's why we're continuing our efforts" (Wager). The report was conceived of as a tool, not just as a publication presenting information. It was information intended to be used. The publication enhanced the chances that activism could succeed because it provided sound information and credibility.

Related Publications: The Extended Rhetorical Situation

Powering the Midwest is still available in 2003, and the executive summary appears on the UCS website, <www.ucsusa.org>. But a new report, *Repowering the Midwest,* published in 2001, has now extended the report. The purpose and methods of the newer report are similar to those of the first report, but the newer report is more directive than the original. Its subtitle, *The Clean Energy Development Plan for the Heartland*, makes clear that the purpose goes beyond reporting research or assessing current energy uses and policy to a plan for action. Like the original report, this one presents data establishing the economic and environmental benefits of changes in energy generation and use. For example, by following the Clean Energy Development Plan, the Midwest can reduce sulfur dioxide pollution by 56%, nitrogen oxide pollution by 71%, and carbon dioxide pollution by 51%. The plan will promote job growth and economic development, while electricity costs will increase only 1.5% (*Repowering* ES-3). These figures represent updates of comparable figures in the first report. The authors are still making their case and persuading the audience of the value of the argument.

This report goes beyond good arguments to plans for implementation, the next step after convincing people in the process of policy change. The need persists to take the data directly to the audience. The specific suggestions are more prescriptive than those in the first report, and they target new audiences, including homeowners. Compact fluorescent lamps will reduce energy use and save Chicago residents $50 per lamp over the lifetime of the lamp, but efficient lighting in business and space cooling produce rapid paybacks. The potential paybacks to industry of using more efficient motors, lighting, and heating depend on the industry: metals fabrication might gain more than agriculture, for example (*Repowering* ES-7). Details such as these recommend particular actions within various settings (homes, business, industry) and justify the definition of the report as a "blueprint."

> *Repowering the Midwest* is a blueprint for producing economically robust and environmentally sound electricity in the 21st century by comparing two possible energy futures for the Midwest—one in which we continue to rely on conventional, or "business-as-usual" technologies, and a second in which the Midwest unleashes its homegrown clean energy development potential. This Clean Energy Development Plan quantifies the region's untapped energy efficiency and renewable resources and lays out strategies, policies and practices to advance a cleaner electricity future from the industrial Midwest across to the Great Plains. These clean power options are technologically and commercially available today, and they can be obtained with only a modest increase in total electricity cost—1.5 percent in 2010 and roughly three percent in 2020—that is far offset by the environmental and public health improvements and the economic and employment gains for our region. (*Repowering* ES-5)

The report targets policy makers as well as homeowners and owners of businesses. One chapter is dedicated to recommendations for policy changes. For example, the authors recommend an energy efficiency investment fund in each state, funded by a charge of 0.3 cents per kWh, to support energy efficiency initiatives. They also recommend energy efficiency standards and building codes and provide details about how to do this. The plan is still a recommendation. "These clean energy resources are now technologically achievable and economically realistic. They will not, however, reach their full potential without significant public policy support" (*Repowering* ES-12–13).

In addition to updates of calculations and a more aggressive plan of action, this report differs from the original as the product of a coalition of organizations. In fact, UCS is second author; the report is primarily the product of the Environmental Law and Policy Center (ELPC), "a Midwest public interest environmental advocacy organization working to achieve cleaner energy resources and implement sustainable energy strategies…" (ELPC). ELPC was founded in 1993 and represents six midwestern states. The lineage to the original report is clear from this statement at the UCS website:

> *Repowering the Midwest* is a collaboration of the Environmental Law and Policy Center, Union of Concerned Scientists, Citizens Action Coalition of Indiana, Iowa RENEW, Izaak Walton League of America, RENEW Wisconsin, Dakota Resource

Council, and Minnesotans for an Energy-Efficient Economy. It is a follow-up analysis to the landmark 1993 UCS study *Powering the Midwest*. (ELPC)

The coalition itself represents progress toward goals and an example of the extended sense of delivery. Getting more people involved in carrying the message to ever wider audiences may be part of what we come to understand as delivery. One organization cannot accomplish alone what eight can accomplish together. Establishment of a coalition is rhetorically strategic in that it expands the number of rhetors and their audiences. The coalition includes citizen organizations from five midwestern states, indicating that UCS and other groups with similar purposes have succeeded in encouraging the development of organized efforts within the states to achieve clean energy goals. The rhetorical ethos of the speaker is strengthened as the organizational support expands. Besides the six organizations listed as sponsors of the study, its authors include consultants in firms with expertise in energy, economics, and the environment. One of these consultants is Michael Brower, lead author of *Powering the Midwest* and now part of Brower and Associates.

Repowering is just one of multiple publications with ties to *Powering the Midwest*. The list of publications in the clean energy section of the UCS website includes new reports almost yearly. Titles include *Energy Security: Solutions to Protect America's Power Supply and Reduce Oil Dependence* (2002); *Clean Energy Blueprint: A Smarter National Energy Policy for Today and the Future* (2001); *Clean Power Surge: Ranking the States* (2000); *Powerful Solutions: Seven Ways to Switch America to Renewable Electricity* (1999); and *Energy Innovations: A Prosperous Path to a Clean Environment* (1997). There are also eleven shorter reports that followed *Powering the Midwest*, as well as others that preceded it. The parent report creates the possibility of additional messages with their own perspective and each tailored to particular problems. Delivery in the sense of carrying messages to the audiences who can act on them does not depend on the single publication. The message is both reiterated and spread, necessary strategies when the goal is change as massive as a shift from coal to renewable resources. The original report, however, is the seed of the others. It represents the research (in terms of the rhetorical canon, invention) and the data that constitutes good reasons.

In addition to printed reports, many of which are available for download as pdf files from the website, UCS uses articles in its newsletter *Nucleus*, press releases, and e-mail letters to members to promote the clean energy agenda. The organization uses multiple channels to communicate its message. One of the main purposes seems to be public education—not just informing readers once but keeping the issue foregrounded. Thus, two *Nucleus* articles about a year apart (Winter 1996–97, Summer 1997) talk about the energy projects in the Midwest. Most readers of *Nucleus* will not have read the reports, so the articles—far shorter, simpler, and less technical than the report—are their only sources of information. *Repowering the*

Midwest is even available as a PowerPoint presentation that website visitors can download for use in their own advocacy activities.

Delivery of the information researched for *Powering the Midwest* does not stop with the initial report or in 1993. Rather the report is part of a series of publications that keep driving the message to the various audiences with power to influence practice and policy. Change of values and of infrastructure cannot happen until citizens and policy makers decide to make them happen. The message will fail if it is conceived to be contained by the single speech or the single publication. Rather, the message is delivered in multiple media by multiple voices over time. Although visual design and typography may capture the original sense of delivery as a visual act, the activities that distribute the information to various audiences over time may better capture the original sense of delivery as an oral act. The visual and oral components give presence and urgency to the ideas. Seeking out the audience rather than waiting for the audience to read a document dramatically increases the chances that the content will be understood and used. The rhetors are more engaged with the audience than the publication model of delivery presumes.

ANALYSIS AND SIGNIFICANCE

The example of the report from the Union of Concerned Scientists confirms the value of the delivery in the sense of visual design and medium. Delivery means accuracy and integrity of the information, representation of information graphically to clarify and not distort information, visual signals to make information accessible and comprehensible and to motivate readers, and the ethos of the author. It also means choice of medium. In the UCS example, the printed report is portable in field work and inexpensive to share with the decision makers who must refer to it, but the website provides additional information and may reach a wider audience, and the newsletter stories and press releases also contribute to making the message available. The follow-up report keeps the information current and the message in the mind of policy makers.

Examination of practice suggests a complex idea of delivery that includes visual design and medium of the publication but also looks beyond the publication to its uses as a tool of strategic action. Delivery is an active address to a complex audience that does not reside in one place at one time. The rhetor is interested not just in persuading the audience of the merits of the argument but also in the actions members of the audience might undertake to implement. This expanded sense of delivery is especially significant to the documents of civic engagement, especially the reports that address exigencies and look for ways to resolve them. Table 1 summarizes some features of reports understood through two perspectives on delivery: the publication concept and the social action concept.

TABLE 1
Comparison of the Publication and Social Action
Concepts of Delivery for Reports
in Environmental Advocacy

Publication	Social Action
Publication is the goal	Publication is the means to an end
Completed study	Continuing activity
Archive (what happened)	Action (what could happen)
Self-contained	Connected, social
Static, regulatory	Strategic, change-oriented
Page design, medium	Engagement
Critique	Practice

In academia, the field of writing studies, including composition and technical communication, has expanded its focus in the past several decades from the text alone to the uses of the text. A correct and accurate text is valued, but it is not enough. A user manual must be usable; a proposal must win support; a website must make searching methods transparent and intuitive to people who visit. Likewise, a technical report, at least one in environmental advocacy, has vitality well beyond its covers. These uses of the documents that technical communication academics teach suggest that the formal qualities of the genres are less an end in themselves than a means of enabling and encouraging their use. This is the social action concept of genre that Carolyn Miller articulated so well for communication studies two decades ago. A concept of delivery that extends beyond publication may help students and professional writers get to this understanding of the relationship of texts to social action. This understanding, in turn, opens possibilities for extended roles for them in civic engagement, not just as editors and grant writers but also as leaders and strategists who can imagine the information needed to determine a good course of action and envision the strategies necessary to persuade decision makers. In short, it helps to develop our students into rhetoricians, responsible not just for accurate texts but also for promoting good public policy and for good outcomes from their work in publication.

This expanded sense of delivery may go beyond what classical rhetoricians had in mind, even as adjustments in theory are made to accommodate written discourse. However, theory adjusts to time and place as well as to medium. In ancient Greece, where the polis was comparatively small, the single speech to an audience might reach the people who could influence change. Even then, however, it is likely that a series of speeches, responding to counterarguments, would have been necessary. Now, there are worldwide audiences for some issues and complex and distant audiences for many issues. People resist change, and there are now more changes to resist than there were in ancient Greece. The increasing complexity of the rhetorical exigence, the audience, and competing arguments on this worldwide stage all argue for a concept of the canons that can expand beyond the single speech bounded by time and place.

The theory of the canons is robust enough to accommodate this expanded sense of delivery. It redeems a trivialized canon by affirming its value in the work of rhetoric in resolving exigencies. Furthermore, a broad definition invites an increased sense of ethical responsibility for the publication and its uses. Even if the writer is not personally responsible for the outreach and use implied by this broad definition, awareness of the uses of discourse in policy and practice encourages students and writers to look outward beyond the page and to aim for something even more significant than accurate representation and effective typography. This sense of delivery also can challenge the suspicion of delivery as distortion. The activities that accompany publication ensure that good ideas reach an audience and are persuasive. Delivery can be a vehicle of understanding and justice.

It is important not to confuse the delivery described here with the spin of political parties nor with marketing of a for-profit organization. These differences may distinguish related efforts: the outreach of an environmental advocacy organization represents an alternative voice, not the hegemonic voice of business or government as usual; the motive is not profit but rather the resolution of an exigence and the public good. The exigence is, of course, defined by words and subject to interpretation and negotiation, but good reasons explain why policy and practice must change and offer alternatives to current practice. Negotiation of public policy is, after all, the business of rhetoric. Technical communication enters into the public sphere as a rhetorical act when the policy issues are technical, as they are in energy policy.

CONCLUSION

The example of the report from UCS is an example of practice that invites a reconsideration of rhetorical theory, especially the theory of the canons, and a reconsideration of the way in which technical reports are taught. The motive for this reconsideration is the potential of our students to develop their public voices and to engage fully in civic affairs, whether as citizens or as professionals. Professional writers continue to be responsible for accurate and usable texts, but at the very least they need to understand the consequences of those texts for immediate and extended audiences, for the present and for a possible future, for other related texts, and for the values that constitute a society. The rhetoric of civic engagement insists, through repetition of message and variation of genre and media, on audience participation. The message cannot remain bound by the document. If one document is limited in persuading audiences to act, another document, another genre, another medium, another emphasis, other settings, and collaborators may be enlisted to help reach the goal. These initiatives, of course, assume that the goal itself is sound and that the information used in persuasion has been developed with good methods and good reasons. One reason why policy initiatives may begin with reports is that the report captures the results of research into the problem and possible solutions. It offers the reasons on which persuasion is based.

The following points would be useful to include in instruction about technical reports related to public policy:

1. When a publication is planned and research is launched, plans for dissemination and use should be part of the planning. In fact, the plan for the publication results from some broader goal, such as developing clean energy as a way to reduce dependence on fossil fuels with their political, environmental, and health hazards. The publication is a means to an end rather than an end.

2. Delivery in the sense of information graphics and typography can distort if the writer does not commit to accuracy and integrity or know good principles of representation. Delivery can also be the vehicle by which good ideas reach an audience who can use them for the public good. The writer's responsibility extends to resolving the exigence that calls the publication into being—to civic engagement.

3. Policy change is slow and incremental. A sound plan for solving a problem is not necessarily heeded until the claims are repeated in other documents and other activities. A web of interrelated and intentional actions moves an idea toward fruition and a plan toward implementation. The burden of change cannot rest on a single speech or a single publication.

4. The writer's responsibility may extend beyond publication. The writer is part of a collaborative team and may not be personally responsible for taking the document into the field, but the writer works with an understanding and vision of a publication within the web of related publications and activities.

Instead of erasing delivery from the canons, this perspective puts delivery squarely back into the center of the work of rhetoric, even when the medium is written discourse rather than speech. It affirms that writers' responsibilities include accurate representation but also the consequences of their words in society broadly defined. It opens the possibility of more significant roles for technical communicators in the public sphere than they have if their work is narrowly conceived in terms of publication alone.

ACKNOWLEDGMENT

I am grateful to Jim Dubinsky and two anonymous reviewers for their generosity in reading with engagement and in providing thoughtful and helpful suggestions.

WORKS CITED

Aristotle. *On Rhetoric: A Theory of Civic Discourse.* Trans. George A. Kennedy. New York: Oxford, 1991.

Bitzer, Lloyd. "The Rhetorical Situation." *Philosophy and Rhetoric* 1.1 (1968), 1–14.

Bolter, Jay David. "Hypertext and the Rhetorical Canons." *Rhetorical Memory and Delivery: Classical Concepts for Contemporary Composition and Communication.* Ed. John Frederick Reynolds. Hillsdale, NJ: Lawrence Erlbaum Associates, Inc., 1993, 97–111.

Brockmann, R. John. *Exploding Steamboats, Senate Debates, and Technical Reports: The Convergence of Technology, Politics and Rhetoric in the Steamboat Bill of 1838.* Amityville, NY: Baywood, 2002.

Brower, Michael C., et al. *Powering the Midwest: Renewable Electricity for the Economy and the Environment.* Cambridge, MA: Union of Concerned Scientists, 1993. UCS <www.ucsusa.org> (accessed 3 March 2003).

Carruthers, Mary. *The Book of Memory: A Study of Memory in Medieval Culture.* Cambridge: Cambridge UP, 1990.

Connors, Robert J. "*Actio:* A Rhetoric of Written Delivery (Iteration Two)." *Rhetorical Memory and Delivery: Classical Concepts for Contemporary Composition and Communication.* Ed. John Frederick Reynolds. Hillsdale, NJ: Lawrence Erlbaum Associates, Inc., 1993, 65–77.

Cook, Kelli Cargile. "Writers and Their Maps: The Construction of a GAO Report on Sexual Harassment." *TCQ* 9.1 (2000): 53–76.

Dragga, Sam. "The Ethics of Delivery." *Rhetorical Memory and Delivery: Classical Concepts for Contemporary Composition and Communication.* Ed. John Frederick Reynolds. Hillsdale, NJ: Lawrence Erlbaum Associates, Inc., 1993, 79–95.

Dragga, Sam, and Gwendolyn Gong. *Editing: The Design of Rhetoric.* Amityville, NY: Baywood, 1989.

Environmental Law & Policy Center. Chicago: ELPC, 2003–04. <www.elpc.org/index.phtml> (accessed 3 March 2003).

Kennedy, George A. *Classical Rhetoric and Its Christian and Secular Tradition from Ancient to Modern Times.* 2nd ed. Chapel Hill: U North Carolina P, 1999.

Levison, Lara. "SEEDs of Victory." *Nucleus: The Magazine of the Union of Concerned Scientists* 18.4 (1996–97). UCS <www.ucsusa.org> (accessed 30 Sept. 2003).

Miller, Carolyn R. "Genre as Social Action." *QJS* 70 (1984): 151–67.

Paul, Danette, Davida Charney, and Aimee Kendall. "Moving beyond the Moment: Reception Studies in the Rhetoric of Science." *JBTC* 15.3 (2001): 372–99.

Repowering the Midwest: The Clean Development Plan for the Heartland. Executive Summary. ELPC, 2001 <www.repowermidwest.org/documents.php> (assessed 17 May 2004).

Reynolds, John Frederick, ed. *Rhetorical Memory and Delivery: Classical Concepts for Contemporary Composition and Communication.* Hillsdale, NJ: Lawrence Erlbaum Associates, Inc., 1993.

Rude, Carolyn. "Environmental Policymaking and the Report Genre." *TCQ* 6.1 (1997): 77–90. Rpt. *Technical Communication, Deliberative Rhetoric, and Environmental Discourse: Connections and Directions.* Ed. Nancy W. Coppola and Bill Karis. Westport, CN: Ablex, 2000. 269–83.

Wager, Janet S. "Renewable Energy: The Road Less Traveled." *Nucleus: The Magazine of the Union of Concerned Scientists* 19.3 (1997). 30 Sept. 2003 <www.ucsusa.org>.

Welch, Kathleen E. "Delivery." *Encyclopedia of Rhetoric.* Ed. Thomas O. Sloane. New York: Oxford UP, 2001. 217–20.

—. *Electric Rhetoric: Classical Rhetoric, Oralism, and a New Literacy.* Cambridge: MIT UP, 1999.

—. "Reconfiguring Writing and Delivery." *Rhetorical Memory and Delivery: Classical Concepts for Contemporary Composition and Communication.* Ed. John Frederick Reynolds. Hillsdale, NJ: Lawrence Erlbaum Associates, Inc., 1993. 17–30.

Yates, Frances A. *The Art of Memory.* 1966. Chicago: Chicago UP, 2001.

Carolyn Rude is Professor of English at Virginia Tech, where she teaches in the professional writing program. She has previously published articles on reports for decision making and reports to support environmental advocacy.

TECHNICAL COMMUNICATION QUARTERLY, *13*(3), 289–306

Rearticulating Civic Engagement Through Cultural Studies and Service-Learning

J. Blake Scott
University of Central Florida

Although service-learning has the potential to infuse technical communication pedagogy with civic goals, it can easily be co-opted by a hyperpragmatism that limits ethical critique and civic engagement. Service-learning's component of reflection, in particular, can become an uncritical, narrow invention or project management tool. Integrating cultural studies and service-learning can help position students as critical citizens who produce effective and ethical discourse and who create more inclusive forms of power. Rather than being tacked on, cultural studies approaches should be incorporated into core service-learning assignments.

A growing number of technical communication teachers and students have found service-learning projects to be fulfilling extensions of their real-world writing assignments. As Thomas Deans suggests, service-learning takes sociorhetorical pedagogies to the next logical step, providing students with wider, community-based audiences, contexts, discourse communities, and modes of collaboration (9). Service-learning also provides students opportunities to develop, reflect about, and enact civic responsibility. This emphasis on civic responsibility can be motivating to students, leading them to look beyond their career preparation or their success in the course, and prompting them to engage with others in community problem-solving.

Yet much of service-learning's promise, including its promise of civic engagement, goes unrealized in many technical communication courses. This is largely because such courses are driven by what I call a "hyperpragmatist" ideology and set of institutional practices and structures. The main goal of this ideology is ensuring students' professional success. Although service-learning comes out of a more robust pragmatist tradition, it can be co-opted by a hyperpragmatism that moves past critique, overlooks broader power relations and textual circulation, and narrowly positions students and their praxis.

This article argues that integrating cultural studies methods into a service-learning approach can ward off this co-option. Further, the integration of the two approaches can bring out the best of each, leading to a robust critical pedagogy of civic engagement. In such a hybrid pedagogy, students engage various community stakeholders as partners rather than follow the directives of "clients," and they critique and intervene in community-based practices rather than merely conform to them. Rather than positioning students only as preprofessional or critical consumers, the hybrid pedagogy I propose positions them as what Jim Henry calls "discourse workers" who channel their cultural critiques into discursive efforts to reshape institutional policy (11). The cultural studies concept of articulation refers to the ongoing process by which coherent structures are produced out of linkages of various elements, such as social formations, ideologies, identities, and practices (Hall and Grossberg 53). This article calls for a rearticulation of technical communication pedagogy through the linkage of service-learning and cultural studies.

In what follows, I first review the legacy of hyperpragmatism, arguing that this ideology and set of practices still permeates technical communication pedagogy, limiting the civic dimension of students' work. Next, I explain how service-learning models have the potential to develop students' civic awareness and engagement. The practical challenges of service-learning threaten to squelch its critical and civic potential, however, especially by relegating reflection to an uncritical, tacked-on exercise and by limiting the scope of students' collaboration with others. Then I argue that integrating cultural studies' emphases and strategies into a service-learning pedagogy can enable it to better ward off hyperpragmatism. After overviewing the contributions cultural studies can make to a pedagogy of civic engagement, I discuss more specific ways I have integrated it into my service-learning courses and assignments, particularly as a way to rethink approaches to user engagement.

LIMITATIONS OF HYPERPRAGMATISM

In the United States, the technical communication course developed out of what Teresa Kynell calls a "milieu of utility" that characterized the vocational education of engineers. Although the earliest versions of the course were more current traditional (i.e., formalist) than vocational, the course gradually began to incorporate more practical elements, such as various types of engineering reports. In the 1940s and 1950s, when the course moved from engineering to English and the field of technical communication emerged, curricula began to consider audience and readability more but still emphasized technical forms, to which the technical article and manual were added (Connors 340–341). Driving this utilitarianism, of course, was World War II and the technology boom it created.

Utilitarian, instrumental approaches have continued to flourish, of course, as attested by the growing corporate influence on curricula and research. Some scholars in the field have even argued that rhetorical theory is counterproductive in the technical communication course—counterproductive, that is, to the goal of training students to succeed in industry. Elizabeth Tebeaux, for example, has argued for vocational (and somewhat current traditional) training that teaches students to "document information clearly, correctly, and economically" (822). Patrick Moore has categorically dismissed rhetorical approaches in favor of instrumental ones that more closely resemble industry skills-based training.

As more researchers and teachers turned to social constructionist approaches, technical communication pedagogy took a more fully sociorhetorical turn. Although early process theory was quite compatible with formalist and instrumental approaches, social process theory helped move the field beyond instrumental concerns to an emphasis on praxis or social, rhetorical action. This turn brought us the emphasis on disciplinary and workplace discourse communities and their writing conventions and processes and prompted us to view students' training as enculturation (C. Miller, "Humanistic"). Yet, even these developments have not shifted the dominant vocational orientation of the field. Most research and teaching still focuses on understanding the discursive practices of organizations or disciplines in order to more effectively adhere to them. Within this paradigm, cultural study is mostly limited to the descriptive (i.e., uncritical) study of organizational culture. As Ann Blakeslee explains in the recently published collection *Reshaping Technical Communication*, which largely overlooks cultural critique and civic engagement in its focus on academic-industry collaboration, "Academic researchers often look at and then *describe* the workplace, specifically its practices and genres. Some aim to convey and *emulate* these practices and genres in the classroom" (43; emphasis added).

Thomas Miller and Dale Sullivan have expanded Carolyn Miller's rearticulation of technical communication as social praxis by arguing that this praxis must be located in larger public contexts and must involve the cultivation of phronesis or practical wisdom. As Thomas Miller explains, this cultivation requires ethical deliberation grounded in the shared knowledge of the community (see also Scott, "Sophistic"). As I will explain shortly, service-learning can embody this more critical and civic notion of praxis.

As an ideology and set of practices, hyperpragmatism is primarily concerned with helping students understand and successfully adapt to the writing processes, conventions, and values of disciplinary and workplace discourse communities. It seeks to help students "conform to the practical conventions of writing that their future employers value most" (France 20). In his advocacy of instrumentalist pedagogy, Moore reveals his primary goal as the following: "Students will *profit* from studying instrumental discourse because they will make the transition from college to the marketplace much more easily" ("Rhetorical" 173; emphasis added). Hyperpragmatism's main priority, then, is to prepare students for successful careers. Along with conformity and effectiveness, hyperpragmatism values efficiency (the smoother students' enculturation, the better). In its more extreme forms, hyperpragmatism can be driven by what Steven Katz calls an "ethic of expediency," where expediency becomes a virtue that subsumes other ethical considerations. Katz argues that our capitalist society privileges an ethic of economic expediency by measuring success and happiness monetarily (274).

Consider, as an example of hyperpragmatism's continuing dominance, common notions of usability in our field. As Barbara Mirel points out, most of our approaches to usability can still be captured by John Gould and Clayton Lewis's 1985 definition: "Any system designed for people to use should be easy to learn (and remember), useful, ... contain functions people really need in their work, and be easy and pleasant to use" (300; quoted in Mirel 168). Carol Barnum's more recent book on usability (and the accompanying PowerPoint presentation on the book's website) similarly defines it primarily in terms of effectiveness, efficiency, and user satisfaction. Approaches to user testing and other usability measures are often touted as ways to make the design process more efficient, please users, and save money. Such views of usability are underpinned by what Bradley Dilger calls an ideology of ease, the "culturally constructed desirability of 'making it easy' or being 'at ease'" (2). We can contrast pragmatic views of usability—grounded in efficiency and ease—with the more critical ones offered by Robert Johnson and Michael Salvo. In *User-Centered Technology*, Johnson distinguishes between user-friendly design and user-centered design. Though more efficient in the short run, a user-friendly model of design, Johnson explains, can ultimately disempower users by failing to engage them as co-designers and by keeping them dependent on experts to understand larger systems and deeper functions (28). By inviting users to be partners in technology design from the beginning of the process, a user-centered model produces technology that is more responsive to users, empowering them to adapt the technology to their complex, dynamic contextual needs. Building on Johnson's theory and Scandinavian design theory, Salvo advances the practice of "user participatory design," which similarly involves a "sustained dialogue between user and designer, when noise becomes the material for information feedback rather than a distressing problem to be avoided or solved" (289). Such reconceptions of usability make room for ethical deliberation and civic engagement in a way that their more pragmatic counterparts do not.

In its attention to the individual development of students based on their vocational goals, hyperpragmatism is politically liberal. It also values the liberal goal of consensus, played out in our field's emphasis on collaboration. Hyperpragmatism's values of conformity and expediency are primarily conservative, however. Carl Herndl and Dale Sullivan have pointed out the tendency of technical communication research and pedagogy to reinforce dominant power relations. Even some studies and pedagogies aimed at change define such change in terms of increasing efficiency and productivity (e.g., Spilka, "Influencing"; Bernhardt); they thereby stop short of challenging the politics of standard practices.

Hyperpragmatist pedagogy is attractive in several ways, not the least of which is its accommodation of students' attitudes about and expectations of their education. Our students increasingly measure their education in terms of economic expediency. Students' self-perception as consumers of professional training (rather than, say, civic training) is tied to the larger corporatization of our field and the university. The increasing corporate sponsorship of our field has its plus side; corporate connections, including those through our trade organizations such as STC, help fund research and curricular programs. Jack Bushnell observes that many technical and professional communication programs have become corporate training grounds to reap internships, fellowships, and good-paying jobs for students (175–76). Administrators advocate corporate sponsorship for these very reasons, of course. But the corporatization of technical communication can also work to reinforce hyperpragmatist values and practices that limit students' development as citizen-rhetors.

As others have pointed out before me, the overly pragmatic orientation of the field is problematic on several counts (Herndl; Longo; C. Miller, "What's"; D. Sullivan). Hyperpragmatism's valuing of conformity can work to reinforce dominant power relations and privilege corporate interests. As Charlotte Thralls and Nancy Roundy Blyler argue, "Uncritically importing into the classroom the communication processes and practices of industry reproduces private corporate interest, making students the tools of capitalist ideology" (15). Dominant power relations and corporate interests are disempowering for many, including technical communicators themselves, many of whom are still marginalized in the corporate cultures of "fast" capitalism as information transmitters (Henry). Hyperpragmatism can also move past or co-opt ethical deliberation and critique, partly by pretending to be practical but not political. Although more robust hyperpragmatist pedagogies teach technical communication as praxis, they often limit this to the social communication practices of discrete organizations, disciplines, or professions and stop short of asking students to critique these practices (T. Miller; D. Sullivan). Herndl points out that moving past critique, pragmatic, social constructionist pedagogy also limits "possibilities for dissent, resistance, and revision" (349); students may therefore see little transformative power in their work. A related limitation of hyperpragmatism is its narrow focus on discrete rhetorical situations and the production processes of specific discourse communities. This focus can be a useful starting point, to be sure, but it also overlooks the broader conditions, circulation, and effects of technical communication. These limitations work together to narrowly position students as preprofessionals. Although students are sometimes taught to be citizens of their employers and perhaps the profession, they are too seldom taught to connect their work (and that of their employers) to larger social issues, too seldom pushed to critique the ethical implications of their work on various publics, too seldom encouraged to engage these various publics as audiences and even partners, and too seldom asked to reflect on their noncorporate social responsibilities.

PROMISE AND CO-OPTION OF SERVICE-LEARNING

In response to these limitations (among other reasons) a growing number of technical communication teachers have been turning to service-learning. Service-learning shares the vocational emphasis on experiential, real-world learning. The service-learning models in which students write for or with their community sponsors most clearly embody this emphasis. Service-learning also expands the contexts of real-world learning to community-based, usually nonprofit organizations and their writing exigencies, the ethical stakes of which can be powerful motivators for students. Further, this learning must meet community needs and must be accompanied by structured reflection about civic responsibility. Such reflection is meant to help students gain both civic awareness and a long-term desire to serve their communities. Leigh Henson and Kristene Sutliff hope service-learning will help their technical writing students "ultimately develop reasonable goals for their own roles in improving the world in which they live" (192). In these ways, certainly, service-learning departs from or at least adds to pragmatic goals. Like others, David Sapp and Robbin Crabtree argue that technical communication and service-learning can make an "ideal pairing." Although typical technical communication courses often function as laboratories for professional training, they explain, an added service-learning component could function as a "companion laboratory in citizenship" that prepares students for responsible participation in democratic life (412, 426).

It is important to distinguish between the hyperpragmatism that service-learning is meant to challenge and the American educational philosophy of pragmatism to which service-learning is theoretically indebted. The latter, most fully developed by John Dewey, acknowledged the political dimensions of discourse, valued the interplay of critical reflection and action, was civic rather than narrowly vocational in scope, and had the goal of social reform. Interestingly, these are also common elements of cultural studies approaches, especially the radical pedagogy of Henry Giroux and others. Indeed, cultural critics Lawrence Grossberg and James Carey describe Dewey as an important figure in American cultural studies.

Despite this partly critical heritage, service-learning approaches to technical communication can all too easily slide into hyperpragmatism. Just as current traditional rhetoric co-opted many process approaches to first-year composition pedagogy (Crowley), hyperpragmatism has co-opted service-learning and other sociorhetorical approaches to technical communication. Nora Bacon addresses this danger when she warns against using an apprenticeship model of service-learning, arguing that such a model uncritically assimilates students into organizational discourses ("Building" 607). The co-option of service-learning by hyperpragmatism is partly the result of the powerful institutional and cultural forces that maintain it and partly the result of the unique demands of a service-learning pedagogy.

For students, the pragmatics of developing, producing, and managing service-learning projects can be challenging, to say the least. As Bacon recognizes, service-learning projects often "introduce students to a community agency one week and expect them to write in its voice the next," thereby requiring a more rapid process of enculturation ("Community" 47). Students must analyze both their sponsoring organization and their text's targeted audience. They must learn and adapt to their organization's values and discourse conventions. They must negotiate their projects in collaboration with each other, their instructor, their project sponsor, and sometimes members of their targeted audience. As part of their larger projects, students' work often involves a proposal, progress report(s), and other documents. The complex, time-consuming tasks of a service-learning project thus leave little time for reflection, ethical intervention, or anything else, especially when the project is initiated and completed within a semester.

Students' excitement about gaining valuable real-world experience and working on worthwhile community projects can prevent them from critiquing the discursive practices that they observe and participate in. In my experience, students can get so caught up in fulfilling their duties to the organization and pleasing their project sponsors that they fail to engage their other audiences or consider the ethical implications of their work for these audiences.

Teachers of service-learning, too, must negotiate the challenges of finding (or helping students find) appropriate sponsors, facilitating an array of student projects (often with different sponsors), teaching students about various genres and conventions, and evaluating the rhetorical effectiveness of student work. In the face of such challenges, it can be easy, to focus only on the practical dimensions of service-learning projects. In addition, teachers often face the pressure of industry-minded administrators and students to emphasize the practical, vocational benefits of service-learning. In reflecting on his previous version of service-learning, Jim Dubinsky writes, "I looked at my course in the mirror. When I did, I saw the reflection of someone who had focused far too much on the instrumental sense of being practical by emphasizing experiential learning's advantages in terms of future employability" (69). Dubinsky is certainly not alone. Technical communication scholarship on service-learning often emphasizes the vocational benefits and practical challenges of this pedagogy for students (e.g., Huckin; Matthews and Zimmerman). One way service-learning courses (including my own) privilege the vocational is by encouraging students to view their sponsoring organizations as practice clients whose accommodation is their main concern. Before he adjusted his pedagogy, Dubinsky observed that his students often "talked about *working for* clients rather than *working with* partners" (69). Unless recognized and addressed, this framing of student roles limits their civic engagement by discouraging them from developing more reciprocal relationships with a wider array of community stakeholders.

Hyperpragmatism's potential to co-opt service-learning is perhaps most evident in approaches to reflection—the key means of fostering civic awareness and, for many of us, the most difficult and vexed part of teaching service-learning. In service-learning courses where reflection is a significant component (in some it is not), it is sometimes relegated to record-keeping logs or personal discovery journals that don't facilitate critical thinking. "Journal writing in many service courses may serve the purpose of creating a log or record of experience," Chris Anson explains, "but falls short of encouraging the critical examination of ideas, or the sort of consciousness-raising reflection, that is the mark of highly successful learning" (169). Journal writing that breaks from pragmatism to encourage personal deliberation often problematically prevents students from recognizing the complexity of civic problems and solutions, from moving beyond uncritical empathy, and from gaining critical self-consciousness (Herzberg 59).

Reflection can also be subsumed into pragmatic invention and project management exercises. Students might be asked, for example, to reflect on the ways their values intersect with those of their organization and/or audience in a rhetorical analysis exercise. Or students might analyze the ethos and conventions of their sponsoring organization, as in James Porter's forum analysis, which examines the background, discursive conventions, and other characteristics of a discourse community. Although such activities qualify as reflection in that they help students connect their projects to course goals, they rarely ask students to critique or propose interventions; instead they aim to help students better understand and negotiate their rhetorical tasks. In addition, such reflection activities can be too focused on the production process, failing to consider the subsequent transformations and effects of students' work as it is distributed and taken up.

Hyperpragmatism can also limit reflection by turning it into a project management tool. Some service-learning courses ask students to reflect on their collaboration, for example, or specific project challenges and their responses to these challenges. Such reflection can be incorporated into more formal assignments, such as the progress report. This type of reflection, too, can be devoid of any critical edge and simply serve as just another tool for making the service-learning project run more smoothly and efficiently.

Some teachers of service-learning are more diligent about framing reflection as critical deliberation. Thomas Huckin has his students discuss the social problems addressed by their community sponsors, answering such questions as "Why are these organizations needed?" and "Why do we have such problems in our society?" (58). Sapp and Crabtree similarly encourage their students to "get a sense of the [sponsoring] agency's 'big picture'" and to reflect on the "larger social issues that give rise to the need for service" (426). Even when reflection is more critical, however, it is often tacked on as tertiary to the project, just as many technical communication textbooks and curricula tack on discussions of ethics. In addition, such reflection often takes place toward the end of the project or course, perhaps in a final evaluation report, when it is too late to act on. Students might be asked to deliberate about their documents' strengths and weaknesses or what they learned from the project, but this amounts to little more than an exercise. Such an approach to reflection may encourage students to be critical of ethical problems but not invite them to intervene in them. The practical demands of a service-learning course can override this type of reflection as well.

Bruce Herzberg points out that service-learning does not necessarily "raise questions about social structure, ideology, and social justice" (59). Several scholars, including Matthews and Zimmerman as well as Henson and Sutliff, have argued that service-learning fosters value development and even long-term civic engagement in technical communication students, but they don't explain how this happens, and they present little evidence to support this claim. In prompting his students to critically examine their relationships with community stakeholders, Dubinsky illustrates one way we can encourage students' civic engagement. I fear that most service-learning courses, however, are closer to previous versions of Dubinsky's (and my) course that privileged vocational values and facilitated praxis but not phronesis. Such courses end up providing additional and perhaps more interesting contexts for teaching the sociorhetorical pragmatics of technical communication, but they don't adequately convert these contexts into opportunities for teaching students to recognize, critique, and respond to the ethical implications of their work.

CONTRIBUTIONS OF CULTURAL STUDIES

Like other critical-civic traditions, including classical rhetoric, cultural studies offers heuristics for extending and challenging hyperpragmatist technical communication pedagogy, as a few scholars have begun to argue. Drawing on radical pedagogy, Herndl, for one, has called for approaches that critique the broader relations of power inherent in technical communication and that enable ethical, public action based on this critique. I have adapted Richard Johnson's notion of the cultural circuit to develop a heuristic that helps students critique and respond to the broader effects of their work as it circulates and is transformed (Scott, "Tracking"). Henry's authoethnography assignment asks students to critique the ways workplace cultures (and fast-capitalist culture more generally) position them and also to strategize ways to revise problematic notions of professional writing and writers.

I've increasingly turned to cultural studies in particular, to ensure the critical and civic potency of student's service-learning work. Like service-learning, cultural studies comes out of multiple traditions and can take various forms. Some of these forms bear little resemblance to service-learning and its emphasis on rhetorical production. In his argument for instilling in technical communication students an "aptitude for cultural criticism," Henry critiques some versions of cultural studies for their "insistence on forming students as critical discursive consumers, all the while wholly ignoring their formation as critical discursive producers in any genre other than the academic essay" (10–11). In contrast, other models of cultural studies, such as those developed by Michel Foucault and Stuart Hall, share with service-learning a commitment to both critical awareness and ethical action.

Rather than teasing out the differences of various models, I'll offer the working definition that has guided my pedagogy. As I'm defining it, cultural studies involves an ethical engagement with, critique of, and intervention into the conditions, functions, and effects of value-laden practices (including discursive ones). Also, cultural studies is as much interested in the circulation and rearticulations of cultural discourses as it is in their generic features. In contrast to approaches that stop at critical awareness or that dance around specific ethical imperatives, the main aim of the model I use is to revise problematic practices to create more egalitarian power relations and more widely beneficial effects. If it is to be effective, such intervention requires civic engagement, especially with those whom it is designed to help.

Cultural studies theory and heuristics can be especially useful to service-learning projects, as such projects have the goal, however unfulfilled, of critical reflection and civic engagement. In addition, such projects require work that has explicit ethical implications and often present students with various ethical dilemmas, including how to negotiate organizational politics or how to balance their duties to their instructor, their organization, and their audiences. At the same time, service-learning can enhance a cultural studies pedagogy by providing a compelling context for students' self-critique and ethical action. A recursive loop of reflection and action is a key element of the Deweyan pragmatism on which service-learning is partly based (Deans 31).

Herndl, Scott, Henry, and others suggest several ways that cultural studies can keep service-learning from lapsing into hyperpragmatism. First, a cultural studies approach requires students to assume a critical stance toward technical communication (especially their own) and the ideological systems of power that regulate it. Beyond learning to produce rhetorically effective texts, students learn to critique their texts' broader conditions and effects. Students might critique, for example, the ways their texts' production processes excludes some perspectives and privileges some values at the expense of others. Or they might interrogate spatial and other boundaries maintained by institutional structures.

Ideally, students' critique would come out of their engagement and deliberation with community partners. Students' civic engagement might draw on Salvo's notion of user participatory design or on the similar impulses of the intercultural inquiry developed by Linda Flower, Elenore Long, and Lorraine Higgins. Flower et al. call for a community problem-solving dialogue that invites community stakeholders to be "partners in inquiry" and offer rival perspectives on the problem and its possible solutions. Such a dialogue values the unique perspectives of all participants and avoids a rush to consensus. Students engaged in intercultural inquiry might also enlist community stakeholders in assessing the effects of their work. This assessment could be based on a broader notion of accommodation that includes responsiveness (i.e., sustained, attentive engagement) and long-term empowerment as well as immediate satisfaction.

On a related note, cultural studies can also enhance service-learning by expanding its sometimes narrow foci on immediate rhetorical situations, discrete discourse communities, and production processes to account for the larger cultural conditions and circulation of discourse. Although typical pragmatic heuristics such as the rhetorical triangle and forum analysis can be useful starting points, they can also lead students to overlook other conditions shaping their discourse and its transformations. In their article outlining an activist approach to service-learning, Donna Bickford and Nedra Reynolds similarly advocate helping students "make connections beyond the site of their encounter" (241). The model of the cultural circuit—which prompts students to track the trajectory of their texts' production, subsequent revision, distribution, reception, and incorporation into people's lives—can help students extend their analysis and build in ways to ensure their texts' ethical sustainability (Scott, "Tracking").

Finally, cultural studies-inflected versions of service-learning can help students channel their critical engagement into discourses calling for ethical revision (Herndl 349). Such revision might occur in the students' projects themselves, particularly in their relations with others. Bickford and Reynolds pose what I think should be a central question for students in a service-learning course: How can we enter relationships "in ways that help destabilize hierarchical relations and encourage the formation of more egalitarian structures" (241)? Students might also direct their revisions—perhaps in the form of the "action plans" described by James Porter et al.—to their teachers, sponsoring organizations, and/or other community stakeholders ("Institutional"). Students might begin to create their action plan by engaging in what Patricia Sullivan and James Porter call "advocacy charting," which can help them strategize their action based on their situated relationships with others involved (Sullivan and Porter, *Opening* 183)

These cultural studies extensions of service-learning will likely complicate students' service-learning projects. If we take students' civic development as seriously as their professional development, however, we will welcome such complication. To make the hybrid pedagogy I'm proposing more feasible, we might first acknowledge that most service-learning projects will not solve community problems or radically redistribute power relationships. Students will not always be in positions of influence. This does not mean that we can't encourage students to address problems in modest ways, however. To make more time for expanded critique and engagement, we might encourage students to take on smaller technical communication projects, remembering as Dubinsky does to treat students' conduct or service as a major text of the course (64, 69). Another way to facilitate students' civic engagement is to, as Sapp and Crabtree recommend, "develop long-term relationships with the cosponsoring organizations" and other stakeholders (426). Larger service-learning projects could be broken up over multiple semesters, each set of students building on (and complicating) the work of the previous set. Students' action plans at the end of the course, for example, might be written to future students who will take up the next phase of the project. I now turn to a more sustained discussion of how I've begun to merge the pragmatic and cultural studies elements of my courses.

SPECIFIC INTEGRATIONS

I began to integrate cultural studies into my own service-learning courses, in which students produce technical or professional communication projects for nonprofit community and campus organizations, first by having students respond to critical reflection questions at different points in the semester but especially in a final evaluation report, where I asked them to critique their documents and the process that produced them. Huckin also describes structured, in-class reflection as something that happens mostly at the end of his courses (58), and Louise Rehling has her service-learning students write a final reflection essay not unlike my evaluation report. But I found that tacking on such reflection to the end of the course left critical reflection secondary to practical production and prevented students from acting on this reflection over the course of their projects.

Lately I've begun to experiment with ways to more fully integrate critical reflection, civic engagement, and ethical intervention into my service-learning courses. Following the model that Melody Bowdon and I propose in our textbook, most of my courses require semester-long projects that involve documents produced for a sponsoring organization as well as several invention and project management assignments, including a discourse analysis, a project proposal, a progress report, user testing, and a final report. These assignments, three of which I expand on below, can be revised to provide contexts for cultural critique, engagement, and action, especially with a focus on the ethics of usability.

My recent experimentation has led me to shift from requiring user testing toward the end of the project to having students consider and enact usability measures throughout the course, even before the project proposal. To make this possible, I help students choose organizations and projects that will enable easy access to their audiences and other community stakeholders. Instead of positioning users as passive subjects of testing and relegating their involvement to the end of the project, a cultural studies approach, informed by intercultural inquiry, would engage users as partners from the problem-defining stage onward. In more recent courses, I've asked students to supplement user testing and the user-test report with other mechanisms that engage stakeholders earlier in the project. A group of students producing brochures for the university's Counseling Center formed focus groups with students living in the dorms—one of their primary audiences—at the beginning of the project. The students then continued to work with the focus group throughout the project, seeking its feedback as they developed the texts' design, language, and so forth. I've had other student groups administer questionnaires to potential users and invite community stakeholders to informal proposal planning meetings.

Like many other teachers of service-learning projects, I begin with a project proposal assignment. The main purpose of this assignment in a pragmatic service-learning course is to provide a blueprint of the project for the instructor, organization, and students themselves. To get to the actual projects, students are tempted to rush through the proposal writing process, which causes them to gloss over the problem and quickly decide on a solution and plan for implementing it. Some savvy students might consider their organization's definition of the problem; many, however, will not incorporate alternative perspectives on the problem and the larger conditions that make it possible.

When students form focus groups or other mechanisms for stakeholder involvement early in the course, these stakeholders can help students and the organization define the problem, its significance, and its underlying causes. Flower et al. call this "getting the story behind the story." Instead of waiting until the end of the course, students can at this point deliberate about the larger cultural and institutional conditions that created and that maintain the problems addressed by their projects. In their proposal for an operations manual, a group working with a non-profit agency promoting music education wrote about the lack of fine arts funding in U.S. education and about widespread ignorance of the benefits of music education. A group producing a leadership training video for a local Boys and Girls Club chapter considered the larger need for more community-based youth leadership programs and opportunities. A group producing job search materials for a halfway house for federal inmates connected the inmates' rhetorical situation to the institutional restraints imposed by the Federal Bureau of Prisons and to other obstacles to reentering society successfully.

Students can also ensure the responsiveness of their projects by inviting stakeholders to help them brainstorm possible objectives and solutions as well as assess the desirability and feasibility of these solutions. For example, a group of students producing a website for a campus organization e-mailed a questionnaire to members and potential members asking them to identify what they'd like to see the website include and do. If they do nothing else at this stage, I require student groups to seek feedback about their project objectives from the organization and at least a few of its clients. This feedback can help ensure that these objectives reflect civic as well as academic and organizational goals.

In its emphasis on cultural circulation, a cultural studies approach to the proposal assignment requires students to think through the changes their texts will undergo after they are submitted to the organization; students consider not only the life cycles of their texts but also how these texts will be transformed along their trajectories. The organization will likely revise and distribute the texts before they are taken up by various users, for example. This type of cultural analysis can help students plan ways to ensure the sustainability of their texts. They could make the texts' design and delivery flexible enough to be distributed in multiple ways. In its proposal, the website group mentioned earlier planned to make the website dynamically updateable and include help documentation on the site itself. To enable flexibility in revision and distribution, the group producing the video planned to produce their text in various formats (VHS, mini-DVD) and submit the raw footage.

In *Service-Learning in Technical and Professional Communication*, Bowdon and I outline a progress report assignment that is somewhat informed by cultural studies. Instead of just reporting their progress in an effort to expedite their project, students discuss a major ethical dilemma they've faced, perhaps involving their conflicting values with the organization, instructor, or each other. A cultural studies-inflected progress report might ask students to assess shifts in power dynamics among project participants or to assess the usability (in the fuller ethical sense of that word) of their emerging drafts. How do these drafts interpellate their audiences, they might consider, and to what possible effects? How well do they account for their users' contexts? Critical assessment at this stage enables students to take action and make changes before submitting their final products. After writing their progress reports, the brochure group decided to form another focus group of students living off campus, and the video group took measures to better incorporate the voices and stories of youth leaders.

I call the final cultural studies-inspired assignment that I will discuss here an action plan (following Porter et al.), but it could also be classified as a recommendation report with an activist slant. As mentioned before, I used to have students write an evaluation report to me and their classmates as the final reflection assignment. In this report, students assessed their project's process and products in light of their objectives and the course's and organization's expectations. Although this assignment asked students to evaluate their work, it mostly led them to reflect on how this work could make a difference for the organization and local community, hopefully inspiring them to get involved in future projects. In addition to being mostly celebratory in nature, this assignment seemed like an afterthought, a final "exhale" after the real exertion of the actual project.

Rearticulating the evaluation report as an action plan with a wider audience (including decision-makers at the organization) gave it more purpose and punch. In the new assignment, students turn their critiques into plans of action for the organization, the instructor, and possibly other stakeholders. The action plan gives students a chance to recommend future courses of action to ensure the long-term benefits of their work and to otherwise address the community problem. Students can recommend that the organization revise, distribute, and assess their texts in certain ways. The brochure group suggested distributing their brochures at new student orientation and in the student union, for example. A group producing a website for a social service agency submitted as part of its action plan user testing directions and materials so that the agency could test the website once it was fully online. I've had students help their organizations set up postproduction focus groups as well.

Students can also use the action plan to recommend alternative or supplementary solutions such as new approaches to producing texts. Students who face difficulties involving users in the organization's text production process can recommend changes to this process. As Jeffrey Grabill's study of a local AIDS planning council demonstrates, the clients of some nonprofit and government organizations face substantial barriers to participation. As supplements to their brochures, the group working with the counseling center recommended fact sheets and new pages on the center's website. It also suggested that the center hold "question and answer" sessions with students (and possibly their families) at the beginning of each semester to inform them of the center's services and to seek their input. This group got some of these ideas from their focus group. Like the project proposal, the action plan can be codeveloped by the students' community partners, the product of civic engagement. Grabill's study illustrates this partnership nicely, as his activist recommendations were based on client input via "involvement meetings" and a questionnaire.

ENHANCED CIVIC ENGAGEMENT

The transformed assignments just discussed are certainly not the only avenues for integrating cultural studies and service-learning. After all, one of the hallmarks of cultural studies is its heterogeneity and adaptation to specific problems and contexts. Other promising avenues include the authoethnographies of workplace cultures taught by Henry and the more activist projects proposed by Bickford and Reynolds. Student activism was a major contributor to the formation of service-learning in the United States (Liu). Others have drawn on cultural studies in fashioning new theories of discourse and research methods that hold promise for service-learning pedagogy. We might draw on the methods of institutional critique, for example, to help students critique "institutions as rhetorical systems of decision making that exercise power through the design of space (both material and discursive)" (Porter et al. 621). This critique could, in turn, lead to proposals for changes in organizational or institutional policies, processes, and spaces. Or we might draw on theories of ecocomposition to help students develop a more nuanced and expansive awareness of the impact of their texts on environments that include but are not limited to humans; in this way, ecocomposition can help us develop a posthuman ethics for service-learning. The next big challenge of implementing service-learning is sustainability, and to this ecocomposition promises to make important contributions. Hopefully this article will inspire others in the field to experiment in the intersection of cultural studies and service-learning.

Both service-learning and cultural studies offer us exit ramps off the highway of hyperpragmatism, a highway that bypasses critical reflection and ethical action and that fails to account for the surrounding conditions and relations through which it runs. Taken together, however, service-learning and cultural studies form an even more promising detour. Cultural studies can keep service-learning from merging back onto the hyperpragmatist highway, a tempting move given its practical demands and the institutional forces driving it. Service-learning can ensure that the use of cultural studies doesn't stop with critical awareness, becoming a dead end. Some versions of cultural studies, especially those influenced by structuralism, start and stop with analysis or position students as critical consumers but not the "discourse workers" Henry describes. If guided by an ethic of engagement and responsiveness, service-learning can also combat the tendency of the cultural critic to determine the needs of subjects for them. As Grossberg argues, some cultural studies approaches problematically assume that the critic "already understands the right skills that would enable emancipatory and transformative action" (385). An approach that draws on the best of service-learning and cultural studies has the potential to rearticulate the technical communication course into a training ground for critical citizens who produce effective and ethical discourse and who work to create more inclusive forms of power.

WORKS CITED

Anson, Chris M. "On Reflection: The Role of Logs and Journals in Service-Learning Courses." *Writing the Community: Concepts and Models for Service-Learning in Composition.* Ed. Linda Adler-Kassner, Robert Crooks, and Ann Watters. Washington, DC: American Association for Higher Education and NCTE, 1997. 167–80.

Bacon, Nora. "Building a Swan's Nest for Instruction in Rhetoric." *College Composition and Communication* 51.4 (2000): 589–609.

—. "Community Service Writing: Problems, Challenges, Questions." *Writing the Community: Concepts and Models for Service-Learning in Composition.* Ed. Linda Adler-Kassner, Robert Crooks, and Ann Watters. Washington, DC: American Association for Higher Education and NCTE, 1997. 39–55.

Bernhardt, Stephen A. "Improving Document Review Practices in Pharmaceutical Companies." *JBTC* 17.4 (2003): 439–73.

Bickford, Donna M., and Nedra Reynolds. "Activism and Service-Learning: Reframing Volunteerism as Acts of Dissent." *Pedagogy* 2.2 (2002): 229–52.

Blakeslee, Ann M. "Researching a Common Ground: Exploring the Space Where Academic and Workplace Cultures Meet." *Reshaping Technical Communication: New Directions and Challenges for the 21st Century.* Ed. Barbara Mirel and Rachel Spilka. Mahwah, NJ: Lawrence Erlbaum Associates, Inc., 2002. 41–56.

Bowdon, Melody, and J. Blake Scott. *Service-Learning in Technical and Professional Communication.* New York: Longman, 2003.

Bushnell, Jack. "A Contrary View of the Technical Writing Classroom: Notes Toward Future Discussions." *TCQ* 8.2 (Spring 1999): 175–88.

Carey, James W. "Overcoming Resistance to Cultural Studies." *What Is Cultural Studies?* E. John Storey. London: Arnold, 1996. 61–74.

Connors, Robert J. "The Rise of Technical Writing Instruction in America." *JTWC* 12.4 (1982): 329–51.

Crowley, Sharon. *Composition in the University: Historical and Polemical Essays.* Pittsburgh: U of Pittsburgh P, 1998.

Deans, Thomas. *Writing Partnerships: Service-Learning in Composition.* Urbana, IL: NCTE, 2000.

Dilger, Bradley. "The Ideology of Ease." *Journal of Electronic Publishing* 6.1 (Sept. 2000) <http://www.press.umich.edu/jep/06-01/dilger.html>.

Dubinsky, James M. "Service-Learning as a Path to Virtue: The Ideal Orator in Professional Communication." *Michigan Journal of Community Service Learning* 8 (Spring 2002): 61–74.

Flower, Linda, Elenore Long, and Lorraine Higgins. *Learning to Rival: A Literate Practice for Intercultural Inquiry.* Mahwah, NJ: Erlbaum, 2000.

France, Alan W. *Composition as a Cultural Practice.* Westport, CT: Bergin and Garvey, 1994.

Grabill, Jeffrey T. "Shaping Local HIV/AIDS Services Policy through Activist Research: The Problem of Client Involvement." *TCQ* 9.1 (2000): 29–50.

Grossberg, Lawrence. *Bringing It All Back Home: Essay on Cultural Studies.* Durham, NC: Duke UP, 1997.

Hall, Stuart, and Lawrence Grossberg. "On Postmodernism and Articulation: An Interview with Stuart Hall." *Journal of Communication Inquiry* 10.2 (1986): 45–60.

Henry, Jim. "Writing Workplace Cultures." *CCC Online* 53.2 (2001): <http://www.ncte.org/ccc/2/53.2/henry/article.html>.

Henson, Leigh, and Kristene Sutliff. "A Service Learning Approach to Business and Technical Writing Instruction." *JTWC* 28.2 (1998): 189–205.

Herndl, Carl G. "Teaching Discourse and Reproducing Culture: A Critique of Research and Pedagogy in Professional and Non-Academic Writing." *CCC* 44 (1993): 349–63.

Herzberg, Bruce. "Community Service and Critical Teaching." *CCC* 45.3 (1994): 307–19.

Huckin, Thomas N. "Technical Writing and Community Service." *JBTC* 11.1 (1997): 49–59.

Johnson, Richard. "What Is Cultural Studies Anyway?" *Social Text* 6 (1987): 38–80.

Johnson, Robert R. *User-Centered Technology: A Rhetorical Theory for Computers and Other Mundane Artifacts.* Albany: State U of New York P, 1998.

Katz, Steven B. "The Ethic of Expediency: Classical Rhetoric, Technology, and the Holocaust." *CE* 54.3 (1992): 255–75.

Kynell, Teresa C. *Writing in a Milieu of Utility: The Move to Technical Communication in American Engineering Programs, 1850–1950.* Norwood, NJ: Ablex, 1996.

Liu, Goodwin. "Origins, Evolution, and Progress: Reflections on a Movement." *Metropolitan Universities: An International Forum* 7.1 (1996): 25–38.

Longo, Bernadette. "An Approach for Applying Cultural Study Theory to Technical Writing Research." *TCQ* 7.1 (1998): 53–73.

Matthews, Catherine, and Beverly B. Zimmerman. "Integrating Service Learning and Technical Communication: Benefits and Challenges." *TCQ* 8.4 (1999): 383–404.

Miller, Carolyn R. "A Humanistic Rationale for Technical Writing." *CE* 40.6 (1979): 610–17.

—. "What's Practical About Technical Writing?" *Technical Writing: Theory and Practice.* Ed. Bertie E. Fearing and W. Keats Sparrow. New York: MLA, 1989. 14–24.

Miller, Thomas P. "Treating Professional Writing as Social *Praxis.*" *JAC* 11 (1991): 57–72.

Mirel, Barbara. "Advancing a Vision of Usability." *Reshaping Technical Communication: New Directions and Challenges for the 21st Century.* Ed. Barbara Mirel and Rachel Spilka. Mahwah, NJ: Lawrence Erlbaum Associates, Inc., 2002. 165–87.

Mirel, Barbara, and Rachel Spilka, eds. *Reshaping Technical Communication: New Directions and Challenges for the 21st Century.* Mahwah, NJ: Lawrence Erlbaum Associates, Inc., 2002.

Moore, Patrick. "Rhetorical vs. Instrumental Approaches to Teaching Technical Communication." *Technical Communication* (1997): 163–73.

Porter, James E. *Audience and Rhetoric: An Archaeological Composition of the Discourse Community.* Englewood Cliffs, NJ: Prentice Hall, 1992.

Porter, James E., Patricia Sullivan, Stuart Blythe, Jeffrey T. Grabill, and Libby Miles. "Institutional Critique: A Rhetorical Methodology for Change." *CCC* 51.4 (2000): 610–42.

Rehling, Louise. "Doing Good While Doing Well: Service-Learning Internships." *BCQ* 63.1 (2000): 77–89.

Salvo, Michael J. "Ethics of Engagement: User-Centered Design and Rhetorical Methodology." *TCQ* 10.3 (2001): 273–90.

Sapp, David Alan, and Robbin D. Crabtree. "A Laboratory in Citizenship: Service Learning in the Technical Communication Classroom." *TCQ* 11.4 (2002): 411–31.

Scott, J. Blake. "Sophistic Ethics in the Technical Writing Classroom: Teaching Nomos, Deliberation, and Action." *TCQ* 4.2 (1995): 187–99.

—. "Tracking Rapid HIV Testing Through the Cultural Circuit: Implications for Technical Communication." *JBTC* 18.2 (2004) 198–219.

Spilka, Rachel. "Influencing Workplace Practice: A Challenge for Professional Writing Specialists in Academia." *Writing in the Workplace: New Research Perspectives.* Ed. Rachel Spilka. Carbondale, IL: Southern Illinois UP, 1993. 207–19.

Sullivan, Dale L. "Political-Ethical Implications of Defining Technical Communication as a Practice." *JAC* 10.2 (1990): 375–86.

Sullivan, Patricia, and James E. Porter. *Opening Spaces: Writing Technologies and Critical Research Practices.* Greenwich, CT: Ablex, 1997.

Tebeaux, Elizabeth. "Let's Not Ruin Technical Writing, Too: A Comment on the Essays of Carolyn Miller and Elizabeth Harris." *CE* 41.7 (1980): 822–25.

Thralls, Charlotte and Nancy Roundy Blyler. "The Social Perspective and Professional Communication: Diversity and Directions in Research." *Professional Communication: The Social Perspective.* Ed. Nancy Roundy Blyler and Charlotte Thralls. Newberry Park, CA: Sage, 1993. 3–34.

Blake Scott is Assistant Professor of English at the University of Central Florida, where he teaches courses in technical and professional communication and the rhetorics of science and medicine. He is the author of *Risky Rhetoric: AIDS and the Cultural Practices of HIV Testing* and, with Melody Bowdon, of *Service-Learning in Technical and Professional Communication.* He is currently working on a book about the rhetoric and politics of the transnational pharmaceutical industry.

TECHNICAL COMMUNICATION QUARTERLY, *13*(3), 307–323

Is Professional Writing Relevant?
A Model for Action Research

Dave Clark
University of Wisconsin-Milwaukee

This article argues that engaged "action research" can help professional writing researchers both develop new and interesting collaborative models and help our profession develop a greater relevance to those not reading our journals and attending our
conferences. I outline one particular, localized approach in the hope that our troubles,
struggles, and failures at University of Wisconsin-Milwaukee can help others to develop their own programs and can further our discussion of community engagement.

In a spring 2003 discussion on the ATTW listserv, Mark Zachry initiated a discussion about the "reach" of our research outside the boundaries of our field, asking us
to ponder the kinds of work we've done with "an audience or stakeholder outside
of our field" and to share how that work has been received, cited, and/or used by
non-TC colleagues (Zachry). Respondents listed interdisciplinary collaborations,
detailing their work with other departments and fields; as one example, Zachry
mentioned the willingness of an engineering colleague to cite technical communication scholarship in a jointly written grant proposal. Both the question and the an-
ecdote reflect an anxiety over the relevance of our scholarship to others. As he suggests, "After spending some time examining a citation index and other search
tools, my sense is that our work is not frequently cited by others," which is true;
scholars in professional writing whose work is used by researchers from other
fields are the exception rather than the rule.

Others worry that our research isn't of relevance even to professional writing
practitioners; Stan Dicks notes that the relationship of academics to practitioners is
troubled by the mutual perception that the discourse produced by the other camp is
"laughably absurd:" "Business people make fun of academic obfuscation and political correctness; academics make fun of people in expensive suits saying
'proactive' frequently" (15). I would add that, as you can learn at any STC conference, there is a theory/practice tension between the two groups. Practitioners often
view theory with suspicion, and academics tend to produce theoretical research

that practitioners wish was better attuned to the day-to-day decision-making on their jobs. As a result, as Dicks points out, practitioners tend to rely not on academic studies, but on private businesses that offer software seminars and online document training. These businesses increasingly base their conclusions and recommendations on research coming from fields other than professional writing, relying on cognitive psychology and third-party advertising in selecting and using new production tools.

A certain level of anxiety over our relevance, then, seems reasonable and worth having, and I'm always excited to hear about new interdisciplinary and industry collaborations that indicate a new level of cultural capital being granted to our work. In what follows, I suggest another path to relevance through the use of entrepreneurial models of community engagement. These models can answer many of our relevance questions, driving the development of applied research valued by practitioners, giving students a broad-ranging practical and citizenship experience, and promoting the relevance of our research and discipline outside our departments and journals.

ANXIETY OVER RELEVANCE

Relevance has been and needs to be one of our long-term goals. Stephen Doheny-Farina has argued that mass-market publication is an avenue to greater outside relevance. Others, like Graham Smart, have taken a more ground-up approach, arguing that to become a mature discipline, "technical communication needs the empirical research that would allow us to develop a body of field-specific 'grounded theory'" (1). In doing so he echoes a long-running line of argument in the field that embracing empiricism can help us create arguments that will be more persuasive and useful to others (cf. Charney).

Still, the biggest push in our literature of the last twenty years or so has been for stronger understandings of ideology in our research. Nancy Blyler and Carl Herndl, among many others, have critiqued research in professional writing for its "descriptive and explanatory" focus, arguing that in conducting our research we must strive to avoid uncritically reproducing the ideologies of our research sites by describing rather than interrogating the ways that power structures and controls discourse. The ideological approach now incorporates cultural studies (cf. Longo) and a variety of theoretical tools that seek to move beyond the limitations scholars see in social constructionist explanations. See, for example, Bazerman and Russell's *Writing Selves/Writing Society* collection and Winsor's *Writing Power,* both of which contain work that tackles studies of ideology and power through activity and genre analysis. Bazerman and Russell's activity theory work, in particular, has led scholars to study new contexts and publish in new venues. But the focus of this work is still largely descriptive, creating thick descriptions of the activity in and across "activity systems" and "genre ecologies" that don't attempt to actively facilitate change in those systems. Publishing is still the primary end goal, and as a result the work resides almost entirely in academics; few professional writing practitioners—or the other types of practitioners described in the work—will read or benefit from these studies.

Recent work of some professional writing scholars has sought to bring "action" to our research, to embrace community engagement as a means to give students nonacademic contexts for their work, to promote an active vision of citizenship, to serve the community, and to create a new variety of relevance for our research. This is a relevance that may not gain us citations in sociological journals or in *Nature*, but it leverages our best research tools and strategies to new applications that have the potential to increase the value placed on our work by a wide variety of community agencies. One variety of such work has been service-learning, a major curricular push here at University of Wisconsin-Milwaukee (UWM), where our Institute for Service Learning works to connect the classroom experience with work in Milwaukee nonprofits. As Ellen Cushman argues, however, community service learning in English Studies has been largely a pedagogical and service-based movement ("Sustainable" 41), where too often the self-defined needs of community agencies are secondary to the "wham and bam" of one-time student projects (40). Although faculty members gain a "new sense of purpose in their teaching," Herzberg suggests that service-learning projects frequently are charitable work rather than attempts to educate students about ideology and social change (59), in a country where the lack of critical governmental support necessitates many of the soup kitchens, literacy programs, and community-based nonprofits in which we place our students. Cushman notes the value of student-based approaches to service-learning (what she refers to as the "end-of-the-semester-project model"), suggesting that this focus "makes good sense given the scholarly agendas one often finds in the field of composition and rhetoric" ("Sustainable" 42), but she further argues that student-based service-learning limits the potential role of academic research in creating sustainable, reciprocal community-based programs. She suggests that we follow Brooke Hessler's call to characterize service learning as "applied scholarship" (61), a designation that emphasizes the importance of sustainability, methodology, and agency-defined results.

Similarly, professional writing scholarship has embraced the ideological approach of our contemporary organizational research, connected it with the ethos of service-learning, and has led to some intriguing new strands of research. For example, Jeffrey Grabill's work in "activist research" within local HIV/AIDS Services seeks not to describe or indict, but to show how "professional writing researchers can help shape public policy by understanding policy making as a function of institutionalized rhetorical processes and by using an activist research stance to help generate the knowledge necessary to intervene" (29). Grabill embraces the ideological standpoint of our organizational research, but suggests through the building of his activist model that we can change as well as problematize these organizations. Similarly, Sullivan and Porter argue in *Opening Spaces* that rhetorical work should be focused on social change (39), and Porter et al. argue in "Institutional Critique" that institutions are "rhetorical entities" and that through rhetoric we can promote change in these institutions (610).

In doing so, Porter et al. acknowledge their debt to research in rhetoric and composition, which has emphasized the importance of change in the mission of rhetorical research (cf. Kirsch; Cushman, *Struggle*). Cushman, for one, has a "stance...composed of three general acts: (1) transferring status, (2) reciprocity, and (3) collaborative knowledge construction" (quoted in Grabill 34), in other words engaging participants in research projects in ways that involve and benefit them as much as the researcher. In professional writing, Jim Henry's work also takes up this thread; in *Writing Workplace Cultures*, Henry focused on transferring status and on collaborative knowledge construction, asking his students to participate in "informed intersubjective research" based on their own questions and interests. "Most writers studied their own workplace, composing 'auto-ethnographies' that problematize these workplaces' local cultures even as they depict writing practices within them." ("Writing"). From his website, Henry also invites future collaborations that enhance reciprocity, calling for interested organizations to join him in joint research projects that will result in "applying," "evaluating," and "designing" work that will directly benefit the organization. Similarly, Brenton Faber argues for "action research" in professional writing, drawing from earlier action research (which "suffered a decline because of [its] association with 'radical political activism' in the 1960s" (188)) to create models that, as Cushman advocates, "seek to improve the quality of people's lives by fully engaging subjects 'as equal and full participants in the research process.' This research begins with an interest in the problems of a community group, and its purpose is to help people understand and resolve the social problems that they face" (189).

Faber concludes his discussion of action research by emphasizing the problems this kind of research can pose for researchers: it begins with a different premise than conventional research, as it is dedicated to helping rather than describing for publication. It also adds new wrinkles to the common research complications of participant observation, data collection, and—as he notes in a discussion of the same Doheny-Farina ATTW speech I just mentioned—the difficulties of employing and articulating theory in work intended to be read and used by those who don't share our background and expertise. These are issues we will spend at least the next ten years working out in our scholarship and day-to-day work. Still, in what follows, I rely on the guidance of Grabill, Sullivan, Porter, and Faber, among others, to articulate one possible model of action research that I hope can promote the application of our research to community problems, develop new applications of pedagogy that can avoid some of the harsher charges against service-learning, and create a new relevance and value for our research in our communities.

BACKGROUNDS AND METHODS

My goal in this article is to suggest an alternative model, a structured approach that can lead to the systematic generation of quality, application-based research that can help our students gain skills and organizational awareness beyond the possibilities of service learning and can improve the perceived relevance of our work among community audiences.

The model I'm going to discuss is in many ways a logical extension of the UWM's in-place model of action research. Several years ago, our recently departed chancellor initiated a program called "The Milwaukee Idea" that emphasizes the importance of dedicating research and pedagogy to community engagement; the Milwaukee Idea seeks to connect scholarly work and service-learning to solve community problems. Through the Milwaukee Idea and its offshoots, the university has committed resources to a diverse range of projects, including building resources for parents struggling to decipher Milwaukee's school choice program, providing technology assessments for nonprofit organizations, founding an Institute for Service Learning that helps align and coordinate student projects and internships, and developing a "Cultures and Communities" program, which was originated by an English department faculty member, that works to promote "diversity and cross-cultural literacy, community-based learning, multicultural arts, global studies, and the cultural contexts of science and health care" ("Cultures").

I was fortunate enough to be a naive new faculty member, unaware of the larger politics, controversies, and resource issues that of course surround the Milwaukee Idea. So I called the vice chancellor and asked for money to start a project, and as my project happily targeted the kind of broad-ranging program advocated by the chancellor, I was granted office space, computer equipment, and enough funds to pay two graduate student employees to assist in my work.

I proposed an active research model that seeks to expand on service-learning models by creating collaborative research engagements with local agencies. In these engagements we rely on qualitative research methods to gather information and follow our data gathering with a holistic, collaborative assessment of an organization's needs and the recommendation and implementation of key projects the organization should undertake. To be more specific, we focus on needs assessment, followed up with the communications training and documents needed for implementation of key, collaboratively designed and organization-selected projects. In the "needs assessment," we work with organizations to produce a report detailing their communications needs. The report is based on quantitative and qualitative research gathered through face-to-face discussion, survey gathering, job shadowing, interviewing, and other techniques, and analyzes current structure and resources, details strategic communication objectives, and proposes a course of action for meeting those objectives. After drafting the report, we enter the "implementation" phase, in which the organization selects, from the array of needs identified in the assessment, a single communication project—for example, a website, a content management system, a grant proposal—for collaborative development and training. They work with us to prepare the document(s), in the process receiving the site- and context-specific training critical for future self-sufficiency. Following the first project, they have a number of options for completing the other projects identified during needs assessment: requesting additional communication projects and paying for them on a sliding-scale basis, internship placements, or more traditional service-learning projects. Beyond that, the organization has a clearly defined and articulated set of needs they can use to acquire funding or other assistance from other providers.

After I offered basic training in ethnographic methods—including interviewing, field notes, triangulation, data analysis, and ethics—to the project's graduate students, hired from within the professional writing program, we developed our ethical stance. We needed to balance our responsibilities to our participants in terms of reciprocity, representation, and collaboration with our needs for research data and experience. To date, we have operated under the following four assumptions:

- We provide a baseline structure to the engagement. As much as I might like to develop a full-blown communication study with a nonprofit, few organizations are willing or interested in designing and participating in such studies. What gets us in the door, as with almost any organizational study, is the perception our work might be useful, and our structuring documents help build that perception. Each potential organization receives a document that summarizes our philosophy, our goals, and the basic structure of the engagement—"during the first visit we'll have you sign consent forms and we'll conduct start-up interviews, so please have the following ready, and so on"—a structure we're still developing based on participant responses.
- We collaboratively design methodologies and objectives to match each case. The purpose of initial interviews is to determine the organization's goals for the engagement; those goals then become the goals of the project. For example, in the case I discuss following, after conferring with the organization we used a combination of ethnographic and management methods, deferred to their hesitation to have us interview their clients and funders, wrote a report, and designed an implementation that targeted their expressed needs.
- We generate a report and then work with the organization to revise it. The report is not jointly generated; organizations value the reports, but our very presence in the organization is often beneficial due to the lack of available resources in an organization, and so far none has had time or interest in generating reports. Assessment and help is, after all, why they allowed us in to begin. So we write the reports; revise to incorporate their feedback to ensure we're targeting the needs we've jointly identified; and lather, rinse, and repeat until we're both satisfied. This ap-

proach, however, worries me. Many readers asked for revision ideas marked grammar problems and declare the document done. But so far our collaborators have seen their self-interest in the reports and offered fruitful suggestions. Although I will certainly use these reports in upcoming articles, the purpose of the report is to serve the organizations. In one case, in fact, the organization was completely uninterested in our report and wanted only the implementation phase of the project, and we agreed.

- We collaborate on all implementations to ensure representation and self-sufficiency. Our claimed areas of expertise include web design and maintenance, knowledge and database management, publicity materials (including posters, brochures, visual identity campaigns), technical documentation, correspondence (including fundraising campaigns), and visual design. In the implementation phase, as in the research phase, we see our responsibility as accurately representing the organization's wishes, providing training—and equipment and software, when necessary and possible, so they can edit and maintain our work after the engagement—and excellent communication.

CASE: THE WSACWC

To date, we have worked with or talked to a wide range of agencies, but we have developed full working arrangements only with three organizations that either came to us directly or to which we were referred by funding agencies. For these pilot programs, we tried to work with a range of different types of organizations. We worked with the Wisconsin State Association of Colored Women's Clubs (WSACWC), an organization that seeks to empower women and children of color; the Wisconsin Women's Business Initiative Corporation (WWBIC), which offers loans, training, and incubation space to minority- and women-owned start-up companies; and Rosalie Manor, which provides a variety of programs for inner-city youth. For our purposes here, I'll focus on the WSACWC, with whom we have had the most interaction to date.

The WSACWC is a historic organization with a national office (the NACWC) in Washington, D.C., and thirty-seven state chapters nationwide, dedicated to serving as an umbrella organization for smaller "member clubs" (like the Phillis Wheatley Club and the Beloit Women's Club) and developing and providing services. Currently, the Wisconsin chapter is working to fund a mentoring program for inner-city youth and is also working to develop a history of its organization. To be more specific, WSACWC organizational mission statement has seven points borrowed from the national organization:

1. To promote the education of women and children;
2. To raise the standards of the home;

3. To improve conditions for family living;
4. To work for the moral, economic, social, and religious welfare of women and children;
5. To protect the rights of women and children;
6. To secure and enforce civil and political rights for the African American race; and
7. To promote interracial understanding so that justice may prevail among all people. ("Legacy")

The WSACWC, which until very recently was operated out of the homes of the club's president and executive director, is currently headquartered in an old inner-city Victorian owned by a member club. The organization uses two upstairs rooms in the house as offices, and the offices are equipped with two computers, a DSL Internet connection, a two-line telephone, a fax machine, a photocopier, and standard office furniture. They still have no funding for their own salaries; they commonly work 60 or more hours a week gratis because of their belief in the organization.

I first met the two women when the vice chancellor referred them to me to help them design a general brochure to promote their organization and their member clubs. When I explained my new project to them, they were enthusiastic about participating. We met and jointly articulated the following objectives that we agreed would help with their long-term missions:

• *Funding acquisition.* A primary goal of the WSACWC is to develop sufficient and reliable funding sources so that they may continue working on their projects and other goals. They cannot continue indefinitely without funding.
• *Club development and membership growth.* The WSACWC would like to develop three additional clubs in politics, business investment and finance, and women's financial issues. Developing these clubs will require hiring a coordinator, someone who understands the organization's vision and focus and can coordinate the programs. In addition, they would like to build the membership of the WSACWC and its member clubs and both provide and more strongly articulate benefits to current and future members.
• *Internal growth.* The WSACWC would benefit greatly from increasing their staff. The organization has already developed an employee handbook and hired a few unpaid interns and is preparing to bring in senior citizen employees provided by local programs. They have developed job descriptions for these positions. This type of assistance will help enormously with the completion of day-to-day tasks, but they also need funds to hire employees with expertise in the areas of their projects.

The president and director estimated that within six months they will undergo drastic changes in the way they operate, especially in terms of paid staff and volunteers. The current focus is on getting funding to operate. They need staff to handle administrative tasks, maintain the facility, and run the programs they are working to establish. In turn, that staff will need to be trained in different areas. After that process, they would be able to focus on building the organization.

After establishing these objectives, and making clear that our goals included gathering data in ways useful to long-term research and student experience, we designed our basic methodology. After discussion, at their request, we did not approach or interview potential funders or clients to assist in our analysis. To reach these objectives, we relied entirely on qualitative methods, including the following activities:

- *Interviews.* For example, during our first formal visit to the organization, we interviewed the president and director in addition to discussing the organization's goals and missions.
- *Field notes.* During our formal visits, we took detailed notes on the organization and diagrammed its physical space, filing systems, and computer setups. We also carefully noted the interactions among the president, director, and the occasional volunteers.
- *Document analysis.* During our visits, we collected documents produced by the organization so that we could analyze their content, organization, and expression.
- *Discussion.* We informally discussed our data among ourselves and with the organization's leaders to create clear findings and recommendations.
- *Ongoing engagement.* We have continued our visits to the WSACWC well beyond our formal visits, visiting once a week on average to work on grant proposals (we have jointly submitted proposals to both SBC and AmeriCorps), to install a computer network, and to chat and learn more about the organization.
- *Collaboration.* The final draft of our report not only reflects the opinion of our staff but has been reviewed, edited, and revised to joint acceptability.

After working with the organization for a while—long enough that we mutually agreed that we understood their organization well enough to attempt drafting a report—I wrote a first draft of the report in which we recommended the projects shown in Table 1 in response to the WSACWC's initial objectives. Items in the "actions" column are repeated to indicate what we see as the cross-objective impacts of this work.

After review, the WSACWC has chosen to have us help them with developing their funding acquisition, a two-action objective that is described as follows (here excerpted from the final report which was prepared at their request with potential funders as the primary audience):

Develop Fund-raising Research, Networks, and Infrastructure

Raising funds is, at base, a communications issue; raising money has everything to do with an organization's ability to effectively present itself and its programs to the right people, at the right times, in the right manner. Few nonprofits can say they

TABLE 1
Recommended Projects for the Wisconsin State Association
of Colored Women's Clubs

Objective	Actions
Develop funding acquisition	Develop fund-raising research, networks, and infrastructure. Expand the advisory board's involvement in external communications.
Club development and membership growth	Expand the advisory board's involvement in external communications. Implement a communication plan. Develop documents needed for public relations, visibility, and interaction.
Internal growth	Expand the advisory board's involvement in external communications. Implement a communication plan.

have all the funding they could possibly need or want. But in the case of the WSACWC, the need for funding is particularly urgent; the two current leaders cannot indefinitely work for free, and ultimately the organization will suffer. In addition, the organization needs funding to pay for outreach (e.g., mailings, a website, events) and for internal needs like supplies and equipment.

We therefore recommend that the WSACWC develop, or have us help them develop, a systematic approach to fundraising that will allow them to fund both their special projects and their day-to-day operation. To be more specific, they need the following, all of which we can either help them develop or point them to better-suited resources:

- Additional grant writing and general fundraising training.
- Improved community visibility for better credibility among potential funders; this should include
  —Comprehensive lists of available funding opportunities.
  —Expertise in searching for new opportunities.
  —Networks of individuals and organizations that can help identify new opportunities.
  —Grant writing infrastructure (templates, editable boilerplate, editorial and writing assistance).

Our participation in this project would be the following:

- Help WSACWC identify training opportunities with area experts.
- Conduct in-house training in writing.
- Create and gather lists of available opportunities.
- Conduct in-house training on searching and applying for new opportunities.
- Help WSACWC identify, organize, and strategize for the growth of their local, regional, and national networking.

- Develop a grant-writing infrastructure (including templates and editable boilerplate).

Expand the Advisory Board's Involvement in External Communications

An organization's board can be a powerful communications tool; most advisory boards are filled with influential community members who have great networking skills and great networks of other influential community members. But at the WSACWC, as at many other small nonprofits, the board's involvement has been reactive rather than proactive—although they respond to specific requests, they rarely act of their own accord to make contacts, network with potential funders, and promote the organization.

The club's president and executive director have limited time and energy, and particularly given their current unpaid status, it is critical that they have 100% networking—and financial—participation from the members of the WSACWC's board. To us, this is a communications issue best solved with better communications. We recommend the following:

- *Board training.* Rather than reinvent the wheel, we recommend the excellent training offered by the Nonprofit Management Fund. This training will help the WSACWC develop strategies for persuading board members to commit to a deeper involvement with the organization.
- *Goal setting.* The WSACWC should compose—formally or informally—a board plan that delineates specific goals for the individual and collective participation they seek from members of their board.
- *Board training for board members.* The WSACWC should meet with its board to renew the board's sense of purpose and communicate their goals for the board's participation.

Our participation in this project would be the following:

- Help WSACWC identify training opportunities with area experts.
- Conduct in-house workshops for goal setting and codifying.
- Prepare training for the WSACWC, likely to be delivered by an area expert in board preparation (this is not currently within the realm of our expertise) (*Communication Assessment* 11–12).

What remains to be seen, of course, are the long-term effects and benefits of our interaction, but the data-gathering and day-to-day work on the project actions have been very rewarding, and each of these subprojects has immersed student employees and me in the minutiae of the funding, politics, and ideologies of our local nonprofit sector.

The engagement we construct through initial research and report generation is the kind of relationship-building work that makes reciprocity and sustainability both natural and practically inevitable, as we combine our fates with the fates of participating organizations. Their failures cannot help but reflect poorly on our collaborations and on the possibility of generating publishable research from our assessments and implementations. Long-term, this work has significant potential to generate multiple benefits for everyone involved, through the generation of student projects (both in the classroom and for master's and PhD projects), the new funding avenues opened through academic/nonacademic partnerships, and the endless possibilities for new work and research; some days the phone calls and e-mail requests are too much to handle. Further, we hope to generate exciting new strands of research possible only through sustained, programmatically supported engagement, including assessing our mounds of data into sectorwide assessments of community needs and similar projects that will allow us to apply and research on macro as well as micro levels.

PROBLEMS AND BENEFITS

It's hard to know where to start describing the problems with our pilot studies. They're too much work and too hard to fit in among the other responsibilities of being a junior faculty member. I am tinkering with models that would allow me to skip, in particular, the laborious process of creating a report. I hesitate to do so only because I value the implicit contract feature of the report, which creates a joint agreement about what the engagement will constitute. I have found the project to be enormously fruitful, generating dozens of new ideas and possibilities, and relentlessly exhausting, consuming as much time as one might wish to devote. In what follows, I detail what I see as some of the problems and benefits with our current approach, in the hope that our problems can be helpful to others.

Conflicting Agendas

As others have noted, organizations and their workers have their own agendas, and if they happen to align with the goals of our research, it's often a coincidence. There are tensions between, on the one hand, our needs to gather research data, create a balanced reward system from the work that gives reciprocal rewards to the organizations, and generate collaborative, multivocal work and, on the other hand, the pressure from nonprofits to narrow our focus and develop self-identified projects as quickly as possible. Virtually every nonprofit has worked with consultants, and the consultant model is both attractive and pervasive. Organizations that have received the reports have valued them as attachable evidence of capacity building and as a detailed, triangulated portrayal of their communication issues, but some, certainly, have indicated that they would prefer that our operation become a free consulting service.

WWBIC's president, for example, wasn't interested in a writing assessment and only wanted our assistance with their website architecture and with an organizational white paper. She considered the time her staff committed to our interviews and other data gathering as the price they paid for our help. And she isn't wrong; there's something to be said for a more direct model. As much as we attempt to make our need assessment reports collaborative, they are still university-produced documents that can be intimidating and therefore unlikely to generate substantive revision comments, although that hasn't seemed to be the case so far. In addition, apart from this article, the reportable work is likely to be generated only by the "implementation" phase of our research, which makes writing the 40+ page reports a time-eating exercise in relative futility.

Funding Realities

Action research is a lot of work. If you don't want to or cannot do it all yourself, it requires a significant amount of external funding, which is both good and bad for our work. All of my previous research, like that of many in our field, has been relatively inexpensive, requiring far less special equipment and resources than, say, a physics laboratory. This is fortunate because professional writing doesn't have the kind of funding infrastructure that supports the sciences, and, as a result, we have tended to engage in smaller, less expensive projects and what my undergraduate students called "nonscientific" work, "opinion-based" articles (like this one) that are unlikely to find audiences outside our profession.

But funding is available, and getting it can be an enormously fruitful part of a research project. Pitching the project to local and regional foundations, to deans, to chancellors, and to nonprofits was absolutely the most critical development phase of the project, as it forced me to learn to articulate academic research goals in terms and methodologies that are useful and meaningful to the Milwaukee nonprofit community, and I received outstanding feedback that has dramatically shaped the project. The money I was granted through the Milwaukee Idea was a one-time gift to the pilot project, and continuing my project has meant learning to build networks of (potentially) interested local funders, a project advisory board, and interested agencies. And the process of learning and building these networks has immersed us in the very communities in which our targeted nonprofits operate; I often find myself giving advice that is drawn less from my research than from advice I myself received only a few days before. Students, too, gain from an exposure to the necessary realities of funding that they wouldn't get otherwise.

Pedagogy

In service-learning projects, students are frequently engaged with organizations on a surface level, working to develop a website or a grant proposal. At best, the student is placed in the organization and thereby gets some in-depth exposure to the workings of the organization. Often, however, a class of students receives a project introduction and a few meetings with their clients, and produces a final product or products. Our fondest hope is that in the process, students will learn new skills, internalize the complexities of working in real contexts, and, through instruction and reflection, come to see their "volunteer" work as the beginnings of a new citizenship for themselves.

Action research, on the other hand, requires deeper understanding of an organization, from inspecting its filing systems to understanding its funding priorities and problems to meeting with its clients and others in its network. Students also get hands-on research experience, filling out IRBs, gathering data, and learning ethnographic ethics and interpretation, and they get paid for their work. Long-term, I see great potential for the project to generate research sites for theses and dissertations that follow either conventional, descriptive research models or engage in action research. As one anonymous peer reviewer of this article insightfully noted, although service-learning is a writing apprenticeship for undergraduates, action research has the potential to be a research and practice apprenticeship for graduate students.

Mingling Research and Service

My department, like many academic departments, does not place significant value on "service" work, whether service is defined as intradepartmental service or as external volunteer or consulting work. I have been warned by the chair, by the vice chancellor, and by colleagues both at UWM and elsewhere that working on such a project prior to tenure is dangerous to my tenure potential. Although colleagues in the department are verbally supportive, there is nonetheless a skepticism about work that doesn't lead directly to publication, preferably in a single-authored book. And there is a definite truth to these arguments. Engagements like these can make publishing difficult as days rapidly fill with a thousand tasks relating to funding acquisition, employee management, and the unstable computer network at the WSACWC. It is far too easy to forget all about publishing as it dawns on you that you are now running your own nonprofit, with stakeholders (board members, community members, etc.) who view publishing as the necessary evil that enables you to keep doing the rest of the work.

But done right, projects like this have the potential to lead to publication; action research need not be unreportable in scholarly forms. I will continue to tinker with the structure of our engagements, streamlining our approach to create an optimal balance between my publishing needs and collaborative research models that can work as service to the organizations. But the potential of the rest of the project is too rich not to tap it; using a service model opens up vast new research possibilities in which organizations are happy to share their time, their documents, their files, and so on with multiple researchers and are happy to collaborate on models of research they view as valuable. I am user-testing a variety of new models with other nonprofits, and the results will fuel my/our research for a long time.

Relevance

Finally, I think that action research, of the type used by my project, can benefit us, our programs, and our universities by bequeathing various kinds of relevance. In my case, our department has received greater visibility on the university level, and I have had opportunities to talk about our theory to at-first skeptical ("You're an English professor?") but ultimately persuaded and interested audiences of funders and community partners, many of whom now come to us. Once relevance to their work is established, the potential projects roll in at a frightening pace, and the trouble is no longer finding research sites but picking and choosing intelligently and sticking firmly to a joint research/service commitment instead of giving up and morphing into a support hotline (not that a hotline is a bad long-term idea).

CONCLUSION

I don't believe that there's any one right approach to action research in professional writing. I hope here only to have introduced some possibilities to inspire further discussion, and I welcome thoughts, comments, and critiques. I suggest that adopting alternative, action research models doesn't have to mean giving up traditional research models. They're just a lot more work, because as a researcher you must produce not only the documents appropriate for our journals and personnel committees but work that is of a more general value to organizations. One of the standard calls that end our academic articles is a call for new work, for adopting new techniques or methods or disciplines in order to expand the possibilities of our work. This article is no exception. I believe that through the process of setting up and running these projects, we can develop new areas of expertise, in funding, in organizational culture, and in management that can help us develop more nuanced, structurally aware analyses in support of our conventional research and our new research partners. Through such projects, we can expose lots of new people and organizations to the power of our research and develop interesting new partnerships (we're exploring linkages with university technology providers and several other Milwaukee Idea units). After the initial shock, all have been willing to concede we had much more to offer than assistance with grammar.

ACKNOWLEDGMENTS

Thanks to Christina Grignon and Patrick Doyle for their work on the project, to the president and executive director of the WSACWC for their collaboration and participation, and to UWM's Center for Urban Initiatives and Research for its financial support.

WORKS CITED

Bazerman, Charles, and Russell, David R. eds. "Writing Selves/Writing Societies: Research from Activity Perspectives". *Perspectives on Writing*, electronic book series. Fort Collins, CO: WAC Clearinghouse and Mind, Culture, and Activity. <http://wac.colostate.edu/books/selves_societies>.

Blyler, Nancy Roundy. "Research as Ideology in Professional Communication." *TCQ* 4.3 (1995): 285–313.

—. "Taking the Political Turn: The Critical Perspective and Research in Professional Communication." *TCQ* 7 (1998): 33–52.

Charney, Davida. "Empiricism Is Not a 4-Letter Word." *CCC* 47.4 (1996): 567–93.

Communication Assessment: Wisconsin State Association of Colored Women's Clubs. Milwaukee: University of Wisconsin-Milwaukee, 2003.

"Cultures and Communities." *University of Wisconsin–Milwaukee* <http://www.uwm.edu/MilwaukeeIdea/CC/main.htm> (accessed 17 Dec. 2003).

Cushman, Ellen. *The Struggle and the Tools: Oral and Literate Strategies in an Inner City Community.* Albany: State U of New York P, 1998.

—. "Sustainable Service Learning Programs." *CCC* 54.1 (2002): 40–65.

Dicks, Stan. "Cultural Impediments to Understanding: Are They Insurmountable?" *Reshaping Technical Communication.* Ed. Barbara Mirel and Rachel Spilka. Mahwah, NJ: Lawrence Erlbaum Associates, Inc., 2002. 13–25.

Doheny-Farina, Stephen. Keynote Address. 2000 Association of Teachers of Technical Writing Conference. Minneapolis, MN. 12 Apr. 2000.

—. *The Wired Neighborhood.* New Haven: Yale UP, 1996.

Faber, Brenton. *Community Action and Organizational Change.* Carbondale: Southern Illinois University Press, 2002.

—. "Intuitive Ethics." *TCQ* 8.2 (1999): 189–202.

Grabill, Jeffrey T. "Shaping Local HIV/AIDS Services Policy through Activist Research: The Problem of Client Involvement." *TCQ* 9.1 (2000): 29–50.

Henry, Jim. *Writing Workplace Cultures: An Archaeology of Professional Writing.* Carbondale: Southern Illinois UP, 2000.

Herndl, Carl. "Teaching Discourse and Reproducing Culture: A Critique of Professional and Nonacademic Writing." *CCC* 44 (1993): 349–63.

Herzberg, Bruce. "Community Service and Critical Teaching." *Writing the Community: Concepts and Models for Service Learning in Composition.* Ed. Linda Adler-Kassner, Robert Crooks, and Ann Watters. Urbana: NCTE, 1997. 57–70.

Hessler, Brooke. "Composing and Institutional Identity." *Language and Learning Across the Disciplines.* 4.3 (2000): 307–19.

Katz, Steven B. "The Ethic of Expediency: Classical Rhetoric, Technology, and the Holocaust." *CE* 54.3 (1992): 255–75.

Kirsch, Gesa. *Ethical Dilemmas in Feminist Research: The Politics of Location, Interpretation, and Publication.* Albany: State U of New York P, 1999.

"A Legacy of Strength." *Women of Vision Young Adult Club* <http://www.expert.cc.purdue.edu/!wov/NACWCHistory.html> (accessed 17 Dec. 2003).

Longo, Bernadette. "An Approach for Applying Cultural Study Theory to Technical Writing Research." *TCQ* 7.1 (1998): 53–73.

Porter, James E., Patricia Sullivan, Stuart Blythe, Jeffrey T. Grabill, and Libby Miles. "Institutional Critique: A Rhetorical Methodology for Change." *CCC* 51.4 (2000): 610–42.

Sullivan, Patricia, and James Porter. "Remapping Curricular Geography: Professional Writing in/and English." *JBTC* 7.4 (1993): 389–422.

Sullivan, Patricia, and James E. Porter. *Opening Spaces: Writing Technologies and Critical Research Practices.* Greenwich, CT: Ablex and Computers and Composition, 1997.

Smart, Graham. "Technical Communication: A Discipline in Need of More Empirical Inquiry." *Proceedings of the 2002 Council for Programs in Technical and Scientific Communication Conference.* Ed. Ann Blakeslee. Logan, UT, 3–5, Oct. 2002.

Winsor, Dorothy. *Writing Power: Communication in an Engineering Center.* Albany: State U of New York P, 2003.

"Writing Workplace Cultures." *Southern Illinois University Press* <http://www.siu.edu/~siupress/titles/00titles/henry_workplace.htm> (accessed 17 Dec. 2003).

Zachry, Mark. "The Reach of Research in Our Field." Online posting. 23 Apr. 2003. ATTW listserv.

Dave Clark is Assistant Professor of English at the University of Wisconsin-Milwaukee. His teaching and research interests include community engagement, the rhetoric of technology and management, knowledge management, and information architecture. His e-mail address is dclark@uwm.edu.

TECHNICAL COMMUNICATION QUARTERLY, *13*(3), 325–340

Technical Communication and the Role of the Public Intellectual: A Community HIV-Prevention Case Study

Melody Bowdon
University of Central Florida

This article argues that technical communicators are uniquely poised to function as public intellectuals. To demonstrate this point, the author offers the example of her work on a major AIDS prevention program report. Situating this work within the history of technical communication, the current discussion of rhetorics of risk, and the writing classroom, the author argues that technical writers don't have simply the opportunity to engage in textual activism; in many cases they have no alternative.

The complicated world in which we live and write and teach demands nothing less of technical communication educators and practitioners than our willingness to be civically engaged as public intellectuals. Economist and legal analyst Richard Posner defines public intellectuals as professionals who draw on diverse knowledge bases to offer authoritative analysis of significant issues of wide concern to nonspecialist audiences. In so doing, they contribute to public understandings of complex issues. Technical communication specialists are uniquely poised to serve as public intellectuals. We have specialized knowledge about ideological dimensions of language that can affect policies and products on institutional and individual levels. We understand historical misconceptions about language as a window pane for reality and can use that awareness to help other professionals to problematize tacit assumptions about audience and other issues that may shape their documents. We recognize the social and political functions of genres and associated conventions and know the importance of that awareness to creating accurate, effective, and ethically sound texts. Our professional and civic engagement can shape the world in which we live and work.

To be public intellectuals, we must recognize our responsibility to (1) make our work part of the public sphere (Mortensen), (2) recognize the democratic functions of our work as educators (Barber; Ehrlich), (3) create positive changes in our communities by recognizing kairotic moments for intervention (Cushman), and (4) recognize our own situatedness within work contexts (Foucault; Haraway). Serving as both a technical communicator and a public intellectual requires that we recognize and embrace the deeply rhetorical nature of our work as shapers of documents, particularly when those documents affect public policy.

To demonstrate this concept of the technical communicator as public intellectual, I will offer an example of my writing and research work on a major HIV/AIDS prevention program report. Situating this work within the history of the field of technical communication, our field's current discussion of rhetorics of risk, and my own writing classroom, I will argue that technical writers don't have simply the opportunity to engage in textual activism; in many cases have no alternative. For this reason, it is worthwhile for our field to continue discussing how we should proceed in complex rhetorical and material circumstances. Further, it is critical that we recognize the importance of seizing kairotic moments for intervention. We must convey to students that they can be responsible rhetorical activists at any stage of document development from composition to publication by collaborating with community stakeholders and specialists in other fields.

This work challenges traditional views of technical communication. In *Writing in a Milieu of Utility*, Teresa Kynell explains that "two very different trends—defense-related writing and the move to a 'humanistic-social' stem in engineering education—provided important linkages in technical writing's distinctive academic place" (75). Like Michel Foucault, Kynell identifies that period immediately after World War II as a critical moment. She points out that the kinds of "frightful and complex" machines that emerged from that war altered the course of engineering practice and training substantially. Foucault argues that those very machines (and their function in what we now call mass destruction) demonstrate that intellectuals must understand the potential uses for any knowledge that they make or help to make; they must situate their work within local contexts to fully understand its power.

Partially in response to the roles of technical communicators in situations such as the well-documented *Challenger* and Three Mile Island disasters, the Association of Teachers of Technical Writing (ATTW) has established a code of ethics that emphasizes the importance of recognizing responsibility and accountability for our uses of language. Some discussion of ethics is included in many of the most popular current textbooks in technical communication, and a few books devoted exclusively to that topic have been published within the past few years by authors such as Dombrowski, Markel, and Allen and Voss. But the model I present goes beyond a focus on ethics in training for technical and professional writing, which implies a responsibility, primarily for honesty in representation of information. My model asks us to cultivate our students and ourselves as public intellectuals whose local action can bring about significant effects in our communities amid current crisis situations such as the AIDS pandemic and other public health problems, as well as the complex world of global politics more generally. We need to do this work consciously and purposefully, and we must see the mediation of truth as one of our required functions as communicators if we are to serve our communities to the best of our ability.

Within the field of rhetoric and composition, scholars like John Trimbur and Richard Miller have encouraged us to see our students and ourselves, respectively, as potential agents of change in academia. In the field of technical and professional communication, Carolyn Miller and Tom Miller address the importance of allowing students to develop as intellectual activists through our instruction and guidance. Both draw on the classical term *phronesis*, often translated as practical wisdom. This kind of wisdom is a major aspect of the role of the technical communicator as public intellectual. Technical communicators must be able to join formal training with real world experience to develop a sense of their responsibility for the rhetoric they produce or help to produce. More examples of technical writers' approaches to dealing with civic responsibility can be found in the special issue of *Technical Communication Quarterly* devoted to the discourse of public policy, edited by Carolyn D. Rude (Winter 2000).

Because, as these and other authors argue, rhetorical issues are almost always ethical issues, in the following sections I'll offer a story from my own experience and demonstrate how my work as a community technical writer and editor required me to push past simple functional definitions of that role to a deeper, more embedded and specific level of civic engagement. My goals here are to demonstrate the kinds of opportunities rhetoricians have to apply their knowledge in local communities; to offer what Foucault calls a "geneology" of a specific project, the "union of erudite knowledge and local memories which allows us to establish a historical knowledge of struggles and to make use of this knowledge tactically today" (83); and to demonstrate the kinds of efforts we may want our students to engage in as they emerge from their university training to serve as potential public intellectuals. Because of our function as liaisons between technical and public audiences and our rhetorical expertise, technical communicators are poised to create change in our local communities and beyond. Scholars like Cushman and Grabill have provided us with models for this work. The following example provides a detailed description of the kinds of engagement we can find in the civic arena and focuses significantly on textual issues that surrounded my work. This case shows that, for me, being a public intellectual is about bringing my specialized knowledge to a community agency, valuing what Haraway calls the situated knowledges of all project stakeholders, and being poised and ready to contribute rhetorical suggestions for shaping the final document.

BACKGROUND: THE GYM (GAY YOUNG MEN) RESEARCH PROJECT

When the AIDS pandemic entered broad public consciousness in America in the mid-1980s, the world was significantly altered. Like the 1940s and 1950s that Kynell and Foucault identify, this was a moment of crisis. This syndrome challenged ideas in public health and scientific ethics. New rules for drug research emerged; new educational standards were created in many communities; and, as Monette, Doty, and others have argued, the lines among medical professionals, patients, and caregivers blurred as knowledge production stemmed from all of these groups. Traditions of top-down health management were challenged in this moment when the medical field was caught off guard and unprepared. These conditions created a fascinating and important site for rhetorical inquiry.

During the late 1990s, I served as a writing consultant on a major research project with the AIDS Prevention Collaboration (APC) of Tucson, Arizona. Using funding provided by the Arizona Department of Health Services and the national Center for Disease Control (CDC), the APC took on a four-year, phase-based, community-level intervention to prevent HIV infection. The first phase of the project was a six-month formative evaluation among key stakeholders—gay young men between the ages of sixteen and twenty-five—called the GYM (Gay Young Men) study. The researchers involved in the project, including a medical anthropologist and a group of trained ethnographers from the public health community, set out to talk with and learn from members of this target population through one-on-one interviews (called "KPIs," Key Participant Interviews) and focus groups. The frameworks for these interviews came from protocols provided by the CDC.

The researchers wanted to find out how young gay men in the area get information about safer sex and what they do with that information. They wanted to find out which methods of conveying safer sex messages get through to the young men and how those messages affect their behavior. They wanted to find out what factors in their subjects' lives affect the ways in which they process such messages. They hoped to use the information they gained to help these young men find ways to assess their own risk-taking behaviors that might lead to infection. They planned to create a matrix model for identifying particular moments and circumstances in young gay men's lives that presented the highest levels of risk of infection. All of this information was to be used to create a set of community-wide strategies for preventing the spread of AIDS through a risk management model.

WORKING THROUGH MISCONCEPTIONS ABOUT TECHNICAL COMMUNICATION: MY WORK IN THE GYM

I became involved in the GYM study project in its final stages. I was called in as a technical writing specialist because the research had yielded some important information that the project team wanted to share with prevention workers in other communities. Unlike Grabill's research with a government AIDS services project in Altanta, where he was involved throughout all stages of a community research project, my role in this project did not appear, on its surface, to be formative. I was asked to edit the report and prepare it for submission to the funding agencies. I agreed, expecting to fix some typos, find and correct overuse of passive voice, and improve the document's format.

Of course nothing is ever that simple, which is one of the concepts we need to help our students understand. As I participated in editing conferences with the authors, I found myself in the more complicated role of public intellectual in the project. In *The Rhetoric of Risk*, Beverly Sauer argues that rhetoricians and technical communication specialists in particular can help field experts to make their risk analysis more useful to the lay audiences whose lives will be affected by them. Jeff Grabill makes a similar argument about the importance of stakeholder involvement in AIDS-related public policy in "Shaping Local HIV/AIDS Services Policy through Activist Research." The issue of stakeholder involvement was central to the GYM project, as the researchers were turning to their target audiences for input on the final products. Whether I liked it or not, I was acting as a public intellectual. I was applying my specialized knowledge to serve as a liaison among groups. I was shaping important documents that would matter to a community I cared about.

When I received the text for the study, I began reading with an editor's eye, but soon I was just reading, listening to the stories of the sixty-one young men between the ages of sixteen and twenty-five whose words were woven into the text amid statistics, literature reviews, descriptions of methodology, and conclusions. What I found as I sat surrounded by medical dictionaries, rhetoric and style manuals, and health journals, was a far more complicated set of problems than the title "editor" had initially meant for me. In this situation I found myself deeply aware of the rhetorical nature of editing that Dragga and Gong, Rude, and others have articulated. I faced serious ethical and rhetorical dilemmas as I struggled to renegotiate my own role in the project. As an AIDS educator and a college instructor, I had a vested interest in the outcome of the report, realizing its potential effects on project funding and the AIDS Prevention Coalition's official prevention strategies in the community. As a rhetorician I could find ways to stand outside the project and evaluate the document that came out of it. As a technical communicator I had conflicting roles. There wasn't a template for negotiating this conflict; the questions I faced went far beyond the parameters of the ATTW professional ethics standards.

I wasn't just editing; I wasn't just generating text; I wasn't just interpreting data. I was doing all of these and more. I knew I had surprising responsibilities, but I didn't have a name for this situation. Was I what Clifford Geertz and others in anthropology call a "participant-observer"? Ultimately I wasn't. I wasn't an epidemiologist, and I didn't participate in the ethnographic research, only in the production of the report that came out of it. Did any of these positionings give me a right to offer substantive recommendations related to the report's conclusions? Facing this question is often a crisis moment for our students working on service-learning and commissioned class projects, because it is linked to issues of authority. This crisis was my impetus for focusing on the model of the public intellectual.

Because I was a language expert called in to assist with clarity and organization in this case, I concluded that my function was unavoidably ideological. As rhetoricians we know that form and content cannot be separated and that the "editing" I engaged in was a highly rhetorical process. My role as a rhetorician and educator was to call attention to the problems I saw in the text and to help negotiate the production of truths in the project. This role placed me in a position of public intellectual, albeit to a relatively small initial "public." In his essay, "Truth and Power," Foucault warns us of the intellectual's complex relationship with the process of producing truths. He writes,

> The essential political problem for the intellectual is not to criticize the ideological contents supposedly linked to science or to ensure that his own scientific practice is accompanied by a correct ideology, but that of ascertaining the possibility of constituting a new politics of truth. The problem is not changing people's consciousness—or what's in their heads—but the political, economic, institutional regime of the production of truth. (133)

I tried to apply this Foucaultian perspective on the production of truth to my work with the GYM study. It was not my job to critique the basic tenets or truths from which the researchers were working, but to remind them of the potential uses their work might have, to remind them to question the "model" they were hoping to design, and to remind them of the possible ramifications such a static piece of "truth" might have in this and other communities to which the report's results might be exported. My outsider status gave me a useful vantage point. As Kathleen Welch suggests, I needed to call attention to the value systems at work in the project. I needed to take advantage of my relationship to the project by raising the other participants' awareness of how these values were being enacted. In this situation, the stakes associated with the truths were high: HIV prevention can be a matter of life and death, and HIV is clearly on the rise among gay communities nationwide. The GYM project was no theoretical, hypothetical, or academic one; it affected our community in major ways. Part of this process of engagement had to do with negotiating my position with the document's authors. Originally they expected me to simply edit the text for correctness and consistency. As we talked in regular editing conferences, they came to appreciate my perspective on complex rhetorical issues and welcomed the shift in my role to public intellectual.

THE STUDY METHODOLOGY

J. Blake Scott provides a thorough analysis of the rhetorics surrounding HIV testing and prevention policies in *Risky Rhetoric: AIDS and the Cultural Practices of HIV Testing*. Scott argues persuasively that the discourse surrounding testing is far more complex than epidemiological models might suggest. Similarly, the subject matter of the GYM study did not involve mathematical formulas or other apparently objective data. The subject was the life stories of young men, and it did not invite easy molding into flowcharts and checklists. Yet that kind of "truth" was necessarily a desired project outcome, as quantifiable data were critical for program assessment and continued project funding.

One of the critical "truths" that shaped the design of this study was a high infection rate among gay men in the community. Epidemiological reports told us that although infection rates in southern Arizona were fairly high, most of the gay men identified as HIV positive were not in the age range being targeted in this study. The GYM project was intended as a prophylactic. In light of this agenda, the project's designers chose to use an ethnographic research approach to accomplish their goals of filling out their image of the young gay man in our community.

The project developers' research suggested that most previous studies of correlations between safer sex campaigns and individuals' behaviors relied on closed models of inquiry that required participants to answer within a range of three to five options. They could respond by indicating that they "agreed," "disagreed," or were "unsure," or perhaps by choosing from options like "never," "sometimes," "often," "usually," and "always." Although drawing on the foundation of this kind of work helped the designers to identify areas of interest that should be considered in their research, this previous methodology did not provide the kind of rich and detailed information that they needed to create an effective prevention. The researchers' decision to take an ethnographic approach to flesh out their knowledge of the "truths" in this community was a cutting-edge move in the context of their field. I believe the nature of the methodology combined with the potential social significance of the results made my role of negotiation in this project particularly difficult.

GAY YOUNG MEN AND THE TROPE OF RISK

Another truth that undergirded the GYM project was the notion that this group of men was at high risk for HIV infection. In *Fatal Advice*, Cindy Patton distinguishes between two paradigms that have informed safe-sex education over the last ten years: "one based on decreasing risk by modifying the behavior of the entire population, and one based on decreasing risk by targeting only those believed to be at highest risk" (23). Clearly the GYM study fell into the latter category. In some sense, then, the participants were always already at high-risk for HIV infection simply because they were part of the gay community. The preliminary version of the GYM report, *Analysis of Gay and Bisexual Young Men's Health Project*, cites research that influenced the shaping of the GYM research plan. The citation includes a list of established factors that place certain young men in the population at particular risk. The second item on the list reads, "They do not fully identify themselves as 'gay' and therefore may not perceive themselves to be in a 'risk group' for AIDS" (*Analysis* 16). The assumption of this statement makes some sense: young men in denial about their sexuality may be generally less honest with themselves and therefore less able to accurately assess their risks. But the statement has another implication. Just being gay or bisexual—not having unprotected anal sex with multiple partners—automatically put these gay young men in a risk group.

Patton explains that this kind of attitude leads to a dangerous essentializing of community members, a move that draws attention away from the kinds of specific simple strategies members of any community can use to avoid infection and directs attention toward a fatalistic image of what it is to be a young gay male in the late twentieth century and early twenty-first century. She refers to the "queer paradigm" in *Inventing AIDS*, suggesting that "straight" and "gay" people who are HIV-positive, regardless of the means through which they contracted the virus, are labeled and understood as "queer" (117). The gay male body is somehow always already infected whether or not the virus enters it or it participates in activities that might create exposure. This always-infected status is accepted as truth despite the reality that there is no biological predisposition to infection present in gay bodies; cultural forces have created the image of inherently endangered and dangerous gay men. Research, science, and prevention education could do little to combat the sense of futility that many of the young men in the GYM study felt. This complex scientific and social dynamic was the backdrop for the GYM project.

Indeed the participants in the GYM project had internalized this message of the inherently at-risk queer body. The *Analysis of Gay and Bisexual Young Men's Health Project* tells us that that "the young gay men in the sample self-reported feeling very vulnerable to HIV and noted that they were very aware that the AIDS epidemic is part of their lives and of their life experiences" (Analysis 17). In fact, 92 percent indicated that they and their friends were at risk for contracting HIV. Although all of the interviewees explained safer sex as (among other things) "using condoms with anal sex," most of them expressed doubts that condoms would protect them from contracting the virus (35).

In spite of their doubt, most of the respondents reported that they used condoms in most of their sexual encounters. Other strategies they used to avoid infection included avoiding sex altogether, avoiding anal sex, and using other partner selection strategies. When asked how they assessed those potential partners, they reported a wide range of indicators, including "perceptions of promiscuity (21%), 'bad' reputation (13%), age (11%) (older men were considered more 'risky'), and physical appearance (18%) (men who look 'sickly' were considered more risky)." They also considered sexually aggressive men and young men who seemed to be in a sexual frenzy to be more dangerous (Analysis 35). Clearly these young men were reporting weak risk assessment models. The goal of this project was to create resources that would help them to improve these strategies.

TEXTUAL INTERVENTIONS

In some ways ethnography has served as a "reality check" in technical writing studies just as the GYM project designers hoped it would for theirs. Our deployment of its strategies has helped us to study rhetoric and to apply it within and beyond academic communities, to identify gaps between our theories of language and its uses beyond the academy. We can also find examples of studies that suggest that ethnography is not the panacea it might appear to be on the surface; this set of strategies brings its own set of potential problems.

As a local and public intellectual involved with this study, I applied fairly simple rhetorical concepts to deal with some problems. In this section, I'll examine two sample problematic passages from the report and explain how basic rhetorical analysis brought them to light and helped to correct them. I must note that no research project could fully capture these narratives and their rhetorical power. No coding system could fully account for the complexities of individual psychology and each respondent's conflicted positions on the issues. I do not present these examples to critique my colleagues' work on the project, but to underscore how useful even the most rudimentary rhetorical interventions can be.

One troubling conclusion of the study came from a brief and terse subsection of a section on behaviors associated with risk-taking tendencies. I'll quote this brief section from the rough draft:

5. Receptive preference
When combined with any of the above predictors, individuals who prefer taking the "bottom" position in anal sex may have a higher proclivity towards risky sexual behavior. (Analysis 37)

This assertion was followed by a segment from an interview:

Informant: On the receiving end, I feel closer to the person. I don't know, but I'm just a true bottom, I guess.
Interviewer: What about condom use?
Informant: I don't want to be bothered. I think that I'm not practicing safe sex with my roommate, um, because he's the person that I normally have sex with, on occasion from these other flings. (Analysis 37)

The report went on to deal with another risk-taking associated behavior. The paucity of support for this generalization concerned me. Scientists and epidemiologists agree that from a clinical biological perspective, the passive (or bottom) party in anal or vaginal sex is more at risk for contracting HIV from an infected partner than the active (or top) party. This makes sense; the passive role creates more physical opportunity for exposure. But that's not what the report says. The report suggests an altogether different point: that the gay young men who identified themselves as receptive in anal intercourse were more affected by cofactors such as child abuse, drug use, transitional phases in their lives, and homelessness. The one quotation used to support this contention had nothing whatsoever to do with the level of impact any of these cofactors had on the interviewee. It was a weak use of evidence to support an important and complex claim.

In my analysis, the claim about the significance of receptive preference supported traditional sexist ideologies about receptive (read here "feminine") parties in all kinds of intercourse—in other words, that passive partners are weaker, less capable of handling pressure, more likely to give in to impulse, and less disciplined. One needs do nothing more than glance at, for example, the history of American medical research in the nineteenth century to see that "science" repeatedly "proved" these very ideas about women with similar kinds of data and support (Schiebinger).

Furthermore, the report seemed to be suggesting that the receptive parties should be targeted for follow-up support; men with a receptive preference should be particularly counseled about safer sex. These suggestions are in perfect accord with traditional attitudes about gender roles and birth control; receptive females are ultimately responsible for taking care of this matter. In *Fatal Advice*, Patton has explored the politics of these approaches in detail, arguing that the ways in which safe-sex campaigns target particular audiences create notions of responsibility for infection. Patton also argues that the campaigns demonstrate the biases and expectations of the educators and administrators who implement them. Could the GYM report be used to justify letting "tops" (young men with nonreceptive preferences) off the hook, in terms of condom use, or targeting them less?

After I presented this issue to the researchers, we were able to go back into the data together and make connections between what the participants reported and the "facts" of epidemiological literature. We were able to present the information in a more concrete and useful way that could be used to help prevention workers identify times when young men might particularly need their help.

Another site in the GYM report troubled me in a similar way. The section of the report, titled "Abuse/Non-Abuse," presents the findings regarding participants' self-reports of sexual and other types of abuse.

> Past events which have caused the individual to have a sense of powerlessness, such as physical, sexual and/or emotional abuse can exacerbate an individual's risk-taking behaviors. When abuse and times of transition are combined, risk for exposure to HIV among gay and bisexual youth is compounded. (35)

Although this conclusion may be valid, the study contained no meta-analysis of possible reasons for the findings. Although several quotations from interviewees who do report sexual and other kinds of abuse from their childhoods are included,

there are no statements from the "nonabused" about their "nonabuse" and its relation to their risk-taking behaviors. There is no exploration of the possibility that the
emergence of a discussion about abuse might trigger a change in the tone of the
open-ended interview toward more trust and personal revelation. Thus, the argument is not complete. I noticed the lack of discussion about possible reasons for the
responses immediately. When I brought up this problem in an editing conference,
we were able to make revisions that made the analysis more useful to prevention
workers.

These revisions, and other similar ones that I encouraged the researchers to
make, were important. Although the unfounded conclusions about the connection
between receptive partners and their histories of abuse might have been extremely
useful for the project members as they planned future prevention activities, this information might also be used erroneously to create the profile of the potentially infected young gay male as a drug-using, suicidal, wild young man who was abused
as a child. This negative profile is certainly not a full and fair depiction of those at
risk for contracting AIDS in any population, and, although it can be useful, it also
may be potentially essentializing. Young men in the gay community of Tucson
who do not fit this narrow profile are not necessarily less at-risk than those who do.
AIDS activist Simon Watney responds to similar problematic models in governmental research:

> It should be perfectly obvious to researchers that the factors determining sexual be-
> havior are not adequately explained by the addiction models which reduce complex
> social phenomena to the status of personal problems. This is victim-blaming of a par-
> ticularly elevated and distasteful kind. There is no such thing as a scientific indicator
> that accurately predicts who is likely to have unsafe sex, nor is there a distinct minor-
> ity of recalcitrant AIDS carriers at large in society who must be detected at all costs
> and forcibly educated. (226)

Watney advocates a global information campaign for young people in every
culture, focusing on reversing the "bad health education" patterns that help HIV
infection to continue to spread (226). Similarly, Patton argues that simplified im-
ages of the "at-risk" person such as that suggested by the GYM report are dangerous (*Fatal Advice* 23). They shift the prevention focus away from actual behaviors
that create risk and onto outward appearance, reputation, and image.

My final story about the GYM report represents a lost opportunity for useful rhetorical intervention, because it suggests that certain key concepts were not adequately developed before the study began. Karen Schriver and others have demon-
strated the importance of document designers' and usability specialists' involvement
early in the process of developing new technologies. I believe that this principle also
applies to the development of studies like the one I was working on in this case. As
I've just discussed, risk and perception of risk were critical concepts for this study,

topics that came up regularly throughout the project. A definition of high-risk (not just "risky") behavior is buried in a dense middle section of the GYM report, and it is much more liberal than the definitions the young men described for themselves in their interviews: "High-risk behavior is defined as three or more incidences of unprotected anal intercourse with casual partners in the previous twelve months with multiple casual partners and drug/alcohol involvement as additional factors" (41).

Although this definition had been established, largely as an adaptation of the definitions put forth in epidemiological journals, I found that the ethnographers involved in the GYM project were not in agreement about this or any other definition of risk. A few weeks into the editing phase of the project, I was in the ethnographer's office going over some notes with her and other staff members who participated in the interviews. She looked up from her work and said to all of us, "How would you define risk?"

A strange conversation ensued. One person mentioned sky-diving at a local tourist spot; another said unprotected sex; one said riding a motorcycle. These people, all of whom had been participating in the project interviews, were defining risk in disparate ways. These were the very people who had worked together to define and code "risk," "risk-takers," "risky behavior," and so forth—the major terms of the study. It was fairly clear (and I followed up with questions to confirm this perception) that in their training sessions for the interviewing stage they had not discussed these definitions. Some might say this was a good thing; perhaps their ideas about risk hadn't been so rigidly codified by the project as to limit their ability to imagine possibilities, but I also wondered how the ambiguity about the meaning of the term *risk* had played out in interviews.

Based on rhetorical analysis of transcripts from interviews, I believe that this disparity yielded both productive and unproductive results. In some interviews the ambiguity of definitions seemed to confuse the participants. In others it seemed to encourage the interviewees to create their own definitions of risk, which helped to enrich the final report and, we hope, the participants' perspectives. Even so, I contend that the results could have been more useful if the interviewers had started with more clear ideas about the meanings of this and other pivotal terms in the survey and about the persuasive nature of language in general. This is the kind of clarification the technical writer acting as a public intellectual can help to shape.

STUDENT INTERVENTIONS

My final and most significant intervention process in this project involved collaboration with my students. Once the final report was written and the results had been reviewed and restated to accommodate the suggestions I (and many others) had made, the model of risk-assessment that emerged was boiled down to a survey that would be distributed among young gay men in the community. The idea was that counselors and social workers could ask the subjects a series of questions that would help everyone involved pinpoint circumstances in which the young men might be likely to make decisions about sex and drugs that might cause them to be exposed to HIV.

I sat down with a colleague from the agency and looked at the survey after the report was written. Although it had finally begun to look useful to us, basically stripped of the kinds of troubling essentialist flags that had worried us during the editing of the report, it still seemed to have a pathological and negative tone to it. As two adults trained in safe sex education, far removed from the experiences of high school and college age kids, we thought it looked pretty good but had a sense that it could use some improvement. We also knew that we weren't equipped to make all of the changes that were necessary. We agreed that I would present it to my students. We would take their suggestions for revisions into consideration.

Because I always tend to bring my writing into my classroom, my technical writing students were aware of my work on this project. Until the day when I brought in the survey, though, what they had mostly heard about was my challenges in understanding certain statistical and medical terms in the report. I had mentioned these experiences in an effort to commiserate with those among them having difficulties with understanding the materials they were working with in their own service-learning projects.

On this particular Monday, though, I brought the project to them in a new way. I began by explaining the background of the study and the purpose of the survey. A couple of students were resistant to working on the project at first, insisting that they couldn't imagine the mindset of anyone who was gay. But soon the entire class was involved and offering excellent and insightful revision suggestions. They were developing new appreciation for the value of user testing and peer review. As a matter of fact, they were giving me back much of the very advice I had been giving them all semester as I commented on their papers: *say what you mean; choose words that are very clear; don't assume your audience will know that; don't be condescending to your readers.*

Because I had explained our concern about the pathological tone of the questions, most of their suggestions were geared toward making the survey seem more upbeat and proactive. They made suggestions about word choices. For example, one question initially read, "Describe your sex life. What's good about, what's bad about it? (Celibacy is considered a sex life!)" After a sophisticated and frequently amusing conversation, they agreed that we should substitute the word "masturbation" for celibacy to make the question clearer and, they believed, more positive. They suggested removing repetitious questions. They recommended changing the order of some questions, particularly in a final section that asked about specific emotional states. They moved a question about suicide to a place in the survey that seemed more logical and more gentle to the reader.

My students completely reworded several questions that they found unclear. First, they thought a question that read "Do drugs and alcohol affect your sex life?" should be changed to "Are you usually sober when you have sexual encounters? If not, what are you on?" This revision took away the values-driven reflection and allowed the respondents to simply report their experiences without the implied judgment of the first version. Another question my students helped to revise was worded "What are the crappiest things that happened to you during your childhood?" Not amused, my students thought a more clear and useful version would be "What good and/or bad experiences from your childhood have made you who you are today?"

My colleague and I agreed that my students were "right." Their suggestions made the document better, and many respondents to the new version of the survey reported positive feelings about completing it. Many got involved in conversations and even programs that have helped them to deal with their life struggles. In their final evaluation of that course, many of my students indicated that that forty-minute segment of our course taught them more about technical writing and its uses and implications than any other experience or assignment in the class. The work we did together helped everyone involved.

THE TECHNICAL COMMUNICATOR AS PUBLIC INTELLECTUAL

The focus of this example has been to describe what I call a dance of ethos—a shifting relationship between the intellectual and her or his community, among multiple layers of individual and disciplinary values. My goal was to describe places and moments in which the rhetorician, acting as a public intellectual, can shape multiple communities. To fulfill that purpose, I'd like to offer some suggestions of specific ways in which technical communication specialists can engage in this work and can model this approach for our students who will have opportunities to shape local, national, and global communities.

First, we can take the information that we've gathered through our research in the history of science and rhetoric to sites where decisions are being made. The people working in the AIDS project I've discussed didn't have time to read about theories of language or persuasion. Underpaid and overworked, they worked hard to keep up with client loads and current research in their own fields. I was able to bring history and theory to them in manageable and relevant pieces. There are several ways we can make use of our knowledge. As Haraway suggests, we can share narratives of the ways "science" has shaped the world we live in the past; we can talk about the kind of impact that social and physical scientists' ideologies have on the ways in which science evolves and then shapes lives. We can highlight the systematic creation of world views in and through science and recommend strategies for seeing, naming, and possibly circumventing the limitations these received worldviews make for us. We can tell stories from the past to those who produce the stories of today.

Second, we can export skills for producing critical frameworks and performing rhetorical analysis to these sites of action. I have focused on rhetorical analysis skills in countless writing class sessions on a variety of levels, but the experience of helping researchers in this field that matters so much to me to create ways to critique texts they read and produce, a parallel teaching experience, was differently satisfying. When I worked with the revisers of this study, we talked about analyzing our basic assumptions, identifying the purposes for our writing; we envisioned various levels of audiences for the texts we were producing and talked about the possible ways people might use the texts for various purposes. We discussed the ramifications of certain word choices and questioned our methods of making conclusions. These issues may all sound obvious to us as writing professionals, but it was helpful for the people I was working with to focus on these issues that they hadn't had time to carefully consider in the past. The third step that rhetoricians can take is to tell the stories of these projects, to let our colleagues and students know about what's going on in these sites, and to identify places that might be good targets for more work.

One of the warnings that Richard Posner gives to his readers in his book, *Public Intellectuals*, is that economic gain is too entrenched a facet of popularly recognized intellectual work in this nation. I believe that one of the best ways to address this concern is through conscientious, ethical, collaborative models of writing instruction. I believe that students need to study potential impacts of their work as writers, engineers, business people, educators, and other leadership roles. They need to recognize the powerful effect for positive or negative change that their work may have in their communities and in the messy world that we all share.

WORKS CITED

Allen, Lori, and Dan Voss. *Ethics in Technical Communication: Shades of Gray.* Hoboken, NJ: John Wiley and Sons, 1997.

Analysis of Gay and Bisexual Young Men's Health Project. Internal Report. Tucson, AZ: AIDS Prevention Collaboration Project, 1995.

Barber, Benjamin. *A Place for Us: How to Make Society Civil and Democracy Strong.* New York: Hill and Wang, 1998.

Cushman, Ellen. "The Public Intellectual, Activist Research, and Service-Learning." *CE* 61.1 (1999): 68–76.

—."Rhetorician as an Agent of Social Change." *CCC* 47.1 (1996): 7–28.

—.*The Struggle and the Tools: Oral and Literate Strategies in an Inner City Community.* Albany: State U of New York P, 1998.

Dombrowski, Paul. *Ethics in Technical Communication.* New York: Longman, 1999.

Doty, Mark. *Heaven's Coast.* New York: HarperCollins, 1997.

Dragga, Sam, and Gwendolyn Gong. *Editing: The Design of Rhetoric.* Amityville, NY: Baywood Publishing, 1989.

Ehrlich, Thomas, ed. *Civic Responsibility and Higher Education.* Phoenix: American Council on Education; Oryx Press, 2000.

Foucault, Michel. *Power/Knowledge: Selected Interviews and Other Writings 1972–1977.* Ed. Colin Gordon. Trans. Colin Gordon et al. New York: Pantheon, 1980.

Geertz, Clifford. *Interpretation of Cultures.* New York: Basic, 1977.

Grabill, Jeffrey T. *Community Literacy Programs and the Politics of Change.* Albany: State U of New York P, 2001.

—. "Shaping Local HIV/AIDS Services Policy through Activist Research: The Problem of Client Involvement." *TCQ* 9.1 (2000): 29–50.

Grabill, Jeffrey T., and Michele Simmons. "Producing Citizens: Toward a Critical Rhetoric of Risk Communication." *TCQ* 7.4 (1998): 415–42.

Haraway, Donna. *Simians, Cyborgs, and Women.* New York: Routledge, 1991.

Kynell, Teresa C. *Writing in a Milieu of Utility: The Move to Technical Communication in American Engineering Programs 1850–1950.* Norwood, NJ: Ablex, 1996.

Markel, Mike. *Ethics in Technical Communication: Critique and Synthesis.* Norwood, NJ: Ablex, 2000.

Miller, Carolyn. "What's Practical About Technical Writing?" *Technical Writing: Theory and Practice.* Ed. Bertie E. Fearing and W. Keats Sparrow. New York: MLA, 1989.

Miller, Richard E. *As If Learning Mattered: Reforming Higher Education.* Ithaca: Cornell UP, 1998.

Miller, Thomas P. "Treating Professional Writing as Social *Praxis.*" *JAC* 11.1 (1991): 57–73.

Monette, Paul. *Borrowed Time.* New York: Avon, 1988.

Patton, Cindy. *Fatal Advice: How Safe-Sex Education Went Wrong.* Durham, NC: Duke UP, 1996.

—. *Inventing AIDS.* New York: Routledge, 1990.

Posner, Richard. *Public Intellectuals: A Study of Decline.* Cambridge: Harvard UP, 2001.

Rude, Carolyn. Special Issue on Public Policy. *TCQ* 9.1, 2000.

—. *Technical Editing.* 2nd ed. Boston: Allyn and Bacon, 2002.

Sauer, Beverly. *The Rhetoric of Risk.* Mahwah, NJ: Lawrence Erlbaum Associates, Inc., 2003.

Schiebinger, Londa. *Nature's Body.* Boston: Beacon Press, 1995.

Schriver, Karen. *Dynamics in Document Design.* New York: John Wiley and Sons, 1997.

Scott, J. Blake. *Risky Rhetoric: AIDS and the Cultural Practices of HIV Testing.* Carbondale: SIU Press, 2003.

Trimbur, John. *The Call to Write.* New York: Longman, 2001.

Watney, Simon. *Practices of Freedom: Selected Writings on HIV/AIDS.* Durham, NC: Duke UP, 1994.

Welch, Kathleen. *The Contemporary Reception of Classical Rhetoric.* Mahwah, NJ: Lawrence Erlbaum Associates, Inc., 1990.

Melody Bowdon is Assistant Professor of English at the University of Central Florida, where she teaches writing courses from the undergraduate to PhD levels and coordinates the Graduate Certificate in Professional Writing. She is coauthor with Blake Scott of *Service-Learning in Technical and Professional Communication*, part of the Allyn and Bacon series on technical communication, and is currently working on a new book project called *Professional Writing in the Nonprofit Sector.*

TECHNICAL COMMUNICATION QUARTERLY, *13*(3), 341–354

Educating "Community Intellectuals": Rhetoric, Moral Philosophy, and Civic Engagement

Michelle F. Eble
East Carolina University

Lynée Lewis Gaillet
Georgia State University

This article encourages technical and professional communication programs to take on the challenge of educating students to become "community intellectuals." The notion of educating future professionals for a career needs to be reconsidered in light of both current research concerning civic rhetoric and past practices in moral humanism courses. The triumvirate of rhetoric, ethics, and moral philosophy provides an effective foundation for reconfiguring existing pedagogy in the field and offers insights for nurturing community intellectuals.

In trying to create projects and assignments that students find accessible, relevant, and practical, we often neglect to provide learning environments that encourage students to develop a civic mindset or move them beyond technological, or what Billie Wahlstrom terms "functional," literacy. Wahlstrom believes that to educate our students to "function as [both] competent communicators and effective citizens, we need to take care to design programs that provide a layered approach to literacy, giving students opportunities to master skills as well as the chance to develop a sense of agency and ethical action" (144). Professional and technical communication courses that emphasize only skill-sets, software, and specific documents fail to consider broader literacies—especially ethical literacies, which potentially can help students become not only effective writers but also engaged citizens who effect change in their communities (Wahlstrom 130).

To prepare students to be both effective writers and engaged citizens, many scholars and teachers now advocate that we design courses that engage with local communities by relying on service-learning and activist research pedagogies

(Bowdon & Scott; Cushman; Deans; Dubinsky; T. Miller). In so doing, we prepare ourselves and our students to consider the "civic purpose of our *positions* in the academy, of what we do with our knowledge, for whom, and by what means" (Cushman, "Rhetorician" 12). Community engagement through service-learning and activist research addresses a need for public intellectuals who work with communities to effect change (Cushman, "Public Intellectual" 328).

The philosophies of the public intellectual have not been fully adopted because of existing negative connotations associated with the term "public intellectual" (Farmer 202). Recent works such as Richard Posner's *Public Intellectuals: A Study of Decline* demonstrate

> an apparent inability to imagine public intellectuals as something other than academics with an identity crisis, as wayward members of the professorate who, after serving apprenticeships in their disciplinary fields, parlay their acquired expertise in the Elysian fields of public advice, debate, and prediction. (203)

To counter this prevailing image of the public intellectual, Farmer offers a new term for the kind of work described by Cushman: "community intellectuals" (204). Farmer's term often circumvents "impoverished understandings of the public as an arena where debates occur among celebrity intellectuals who display their argumentative prowess to no apparent end" (203–04). He suggests that the term "community intellectuals" might "remind us that there are other publics and other intellectuals whose efforts, although often unheralded, make an authentic difference in the lives of our neighbors" (204). The term "community intellectual" also emphasizes the need for professionals to participate in rhetorical engagement within their own local communities, to get their hands dirty (so to speak) through active participation in local communities.

The call to emphasize both practical and humanistic concerns in the teaching of professional communication is not new and is often linked to our historical connections with classical rhetoric and the notion of working for the good of society. In "Treating Professional Writing as Social *Praxis*," Thomas Miller explores connections between professional writing and the tradition of civic rhetoric and argues for a view of professional writing as social practice. He claims that "we cannot be both technicians of the word and humanists because there is a basic contradiction between teaching writing as a technique of information processing and teaching writing as the negotiation of shared values and knowledge" ("Treating" 70). Miller asserts we need to make a choice between the two approaches. One approach has a "long practical philosophical tradition behind it, ... advance[s] humanistic values, [and] is practically engaged in the world beyond the classroom. The other approach has made us...technicians who can help businesses better convey their messages but cannot question how and why those messages have been chosen"

("Treating" 70). His vision of this crossroads in professional communication programs is just now—over a decade later—being realized in professional and technical communication program and course development as we blend concerns of the rhetorical tradition with community engagement. For example, in a recent article linking classical rhetoric, service-learning, and professional communication, James Dubinsky argues that service-learning provides a way to bridge the gap between practical and humanistic concerns, and can serve as "a path toward virtue and can create ideal orators and citizens who put their knowledge and skills to work for the common good" (62).

The work of scholars and teachers like T. Miller and Dubinsky demonstrates that one way for programs in professional and technical communication to balance pragmatic and humanistic concerns and nurture community intellectuals is through a civic philosophy of rhetoric and a pedagogical borrowing from our field's roots. In this article, we examine the work of moral philosophers from the eighteenth century, believing that today's professional and technical communication teachers share many of the concerns confronted by moral philosophy professors of the eighteenth century. Those professors, by blending the study of rhetoric, ethics, and politics with students' interests, helped meet the needs of their students, business/industry, and society. Basing their work on the rich tradition of civic humanism, those teachers and their courses provide instructive models for balancing both humanistic and practical matters in our writing courses and programs.

In this article, we begin by outlining a brief history of the moral philosophy courses during the eighteenth century in Scotland and in early America. These earlier courses in moral philosophy inform present-day technical and professional communication theory and practice and provide both a precedent and validation for encouraging and preparing our students for rhetorical engagement in society and the workplace. Following this discussion, we explore applications of layering literacies in our own professional communication courses in light of historical precedents set by our professional forebears. Then, in the final section of this article, we examine how service-learning can help create community intellectuals—if that pedagogical practice is combined with instruction in rhetoric, ethics, and community engagement.

RHETORIC, MORAL PHILOSOPHY, AND THE TRADITION OF CIVIC HUMANISM

Eighteenth-century moral philosophy courses originated from an adaptation of Ciceronian rhetoric. Noted authority Wilbur Samuel Howell explains that "when eighteenth-century logicians referred to ancient logic, they thought only of Aristotle, whereas their rhetorical colleagues in conceiving of ancient rhetoric were completely unable to think of anyone but Cicero" (75). Cicero's rhetorical theory is directly attributable to and gains its authority from Aristotle. Howell urges us to see "Ciceronianism in rhetoric as the counterpart of Aristotelianism on logic." He further explains that "the two systems are so intertwined that the problems of one are in fact variations upon the problems of the other, and a firm grasp of history of both is essential to the understanding of the fate of either one" (76). Many of the eighteenth-century theorists claim allegiance to Ciceronian rhetoric because of the rhetorician's recognition "that the exchange of ideas between one person and another, or between one generation and another, is at the very center of man's social, political, moral, economic, and cultural life, and that any art which improves man's capacity to exchange ideas is at the very center of all the other arts" (Howell 76–77). Cicero was a powerful opponent of those who threatened the welfare of the Republic. In *De oratore* (55 BC), his major treatise on rhetoric, he discusses the significance of rhetorical engagement for society and outlines the role rhetoric and philosophy play in education and civic affairs. Eighteenth-century Scottish educators and political activists were attracted to Cicero's dual role as both a public advocate for the Republic and a rhetorician. As Howell sees it, by the eighteenth century, Cicero "was counted the greatest writer on rhetoric ever to have achieved first rank both as an orator and man of letters" (76).

During the eighteenth century, Scotland was in a position to adapt Ciceronian concerns for their own needs. The 1707 Act of Union between Scotland and England, which united the parliaments of the two countries, brought great prosperity to Scotland. No longer prohibited from international trade nor taxed in trade with England, the merchant class (particularly in the port city of Glasgow) became quite wealthy and demanded changes in education, heretofore designed primarily to train men for the clergy, which would prepare men for careers in business and industry. Although no longer independent, eighteenth-century Scots kept control over (1) their legal system, based on continental civilian law rather than English common law; (2) the National Church of Scotland, largely Presbyterian and marked by a history of dissension; and (3) their excellent educational system—which included four major universities far superior and more progressive than England's Oxford and Cambridge.

The Scots supported a philosophical belief in a democratic education with few religious restrictions, open to talented male students who were interested in pursuing higher education. Obviously, the Scots' perception of a democratic education doesn't parallel our own—women were not admitted to the four major Scottish universities, although women's academies were in existence. The Scottish educational attitude derives from earlier educational practices in the region. Before the major universities were established in Scotland in the fifteenth century, college students attended universities on the continent—particularly in Paris and Bologna (Hunter 209). Therefore, the model for the founding of the Scottish schools was the broadly democratic institution of the continent rather

than the traditionally elitist English school. The philosophical differences between the English and Scottish universities became important in the eighteenth
century. Winifred Bryan Horner explains,

> In the English universities…only the best prepared and brightest students were ad-
> mitted…while Oxford and Cambridge were concentrating on the classics, the Scot-
> tish universities were embracing the new learning. It is within this atmosphere that
> the study of English literature and critical theory and the new psychological rhetoric
> emerge. (*Nineteenth-Century Scottish Rhetoric* 3–5)

The lasting contributions to liberal arts education made by the Scottish universities are directly attributable to eighteenth-century changes in their philosophical
approach to education. In an effort to better serve their students, the Scottish universities during this period were (1) shifting from regenting/tutorial systems of instruction, (2) moving away from deductive reasoning toward inductive inquiries
into social and physical experience, and (3) incorporating concentrated studies of
English language and logic, and studies of social relations of contemporary life
into the curriculum.

Perhaps the greatest change at the Scottish universities during this period was in
the demographic makeup of the students. The class no longer met the specific
needs of young men seeking to raise their stations in life and compete with English
students for jobs. The Scottish university students were often as young as thirteen
or fourteen in the late eighteenth century and were graduated at age seventeen or
eighteen (Hunter 21; Findlay 9–10). By the nineteenth century, the University of
Glasgow was attracting a diversified range of male students, consisting of many
different "ages, classes, and occupations" (Horner, "Nineteenth-Century Rhetoric" 174). Because these students were younger than their predecessors, they were
often educationally unprepared for the lectures in ontology, metaphysics, and
Greek, which characterized the logic and philosophy class. In addition, they were
lured away from college at an earlier age than students of the past because of increased employment opportunities both in Scotland and abroad, opportunities,
which children of the working classes could not easily afford to ignore. Because of
the shortened time spent at college, the students' education became more prescriptive and abridged. The changes taking place in society required a more miscella-
neous and practical kind of instruction in the required first philosophy class.

MORAL PHILOSOPHY

During the eighteenth century, professors of moral philosophy in the British cultural provinces included the study of English literature, composition, and rhetoric
in their course curricula; all students were required to take these courses. Working

in the margins of the British realm, moral philosophy professors, such as Adam Smith and John Witherspoon, educated their students to compete with Oxford and Cambridge-educated students for jobs and social position. Building on the work of perhaps the most important theorist of the Scottish Enlightenment, Frances Hutcheson (professor of moral philosophy at Glasgow University for sixteen years), these first professors to teach English Studies could be defined as "civic rhetoricians," professors "concerned with the political art of negotiating received beliefs against changing situations to advance shared purposes" (T. Miller, *Formation* 34). In many cases, these professors delivered public lectures in English to citizens interested in social, political, and economic advancement, and later the curriculum of the public lectures found its way into the university courses. Of note, the Universities of Edinburgh and Glasgow were also urban and governed in part by town councils composed of local citizens and merchants who encouraged changes in university curriculum to reflect concerns of business and local industry. Core courses in moral philosophy addressed these interests and concerns, including questions concerning the compatibility of capitalism with traditional Scottish ethics and virtues of social welfare, justice, and shared deep philosophical beliefs.

Moral philosophy classes, particularly in the late-eighteenth-century Scottish universities, concerned rhetoric, economics, and ethics while promoting "a conception of language that emphasized the moral value of the study of aesthetics" (Court, *Institutionalizing* 14). Paving the way for later courses in moral philosophy, Adam Smith, student of Frances Hutcheson and himself Chair of Moral Philosophy and Logic at Glasgow University from 1751 to 1764, modified his novel Edinburgh University course (1748–51) to merge his thoughts on literary criticism and rhetorical theory with his interests in ethics and economics. Franklin Court explains, "The philosophical rationale behind his case for the formal study of English literature was closely connected to his thoughts on 'sympathy' and his argument for the education of the 'good man' (the 'good bourgeois') and 'studious observer'" (*Institutionalizing* 12).

Smith's civic humanism, traced to Cicero's thoughts on civic duty, prompted him to devise a course of study that would eventually link the self-interest of his students with public concerns. Smith believed that the formal study of literary characters' ethical and unethical behavior—based on "close textual examination and interpretation"—would lead students "to share experiences and feelings over time through a process of associative, imitative identification that naturally approved good acts and deplored evil ones" (Court, *Institutionalizing* 12). In particular, Smith was interested in the ethos of speakers/communicators and the rhetorical considerations guiding particular moments of rhetorical engagement. Smith's lectures reveal a clear taxonomy of composition "linking rhetorical theory and practice to cultural contexts" (Ulman, *Encyclopedia* 674). Smith addresses didactic or scientific writing, explaining that the purpose of scientific writing is primarily instruction, not persuasion. The goal of this category of writing, therefore, is often not arguing in a classical sense, but rather it "proposes to put before us the arguments on both sides of the question in their true light" (Smith 62). Smith was enormously influential, and subsequent classes in a variety of disciplines owe homage to his eighteenth-century influence. Through the work of Hugh Blair and George Campbell, Smith's famous successors, Smith's belletristic interests become the "key components of rhetoric for the succeeding century" (Bizzell and Herzberg 651), but it is Smith's student, George Jardine, who codified Smith's rhetorical theory in his detailed description of a unified teaching plan. Jardine's conception of education was utilitarian: to prepare young men for careers in business and science. Jardine was teaching communicative skills to his students to enable them to break class bonds and become competitive in British society. He was attempting to improve the quality of education for his students in the larger interest of creating a more just, radically democratic society. As a result, his pedagogy, although based on a curriculum associated with enlightenment rhetoric, closely resembles our own. Jardine envisioned a comprehensive rhetoric, stressing that the abilities to reason, to investigate, to judge, to write, and to speak are crucial components of a liberal arts education. According to George Davie, "Jardine found imitators everywhere; at St. Andrews, the Professor of Logic, Spalding, gained fame as a second Jardine for his ability to work the students hard, and to inspire them with a serious interest in writing essays on cultural subjects" (19). Jardine expanded, nationalized, and exported cultural and rhetorical concepts he learned from Smith and applied Smith's earlier theories to emerging interests in writing instruction. In turn, Jardine influenced many subsequent educators.

Perhaps the most interesting figure in the eighteenth-century revival of the civic rhetoric tradition (although not the most influential) is John Witherspoon. Born the same year, both Smith (1723–90) and Witherspoon (1723–94) influenced a revival of Ciceronian rhetoric on two continents: Smith in Great Britain, and Witherspoon in America. Smith's interest in civic humanism informed his moral philosophy courses and rhetorical theory, whereas Witherspoon, like Cicero himself, was far more interested in rhetorical engagement than rhetorical theory. According to bibliographical profiles, Witherspoon, educated at Edinburgh University with his classmate Hugh Blair, emigrated to the American colonies in 1768, where he would become president of Princeton University and an influential leader in the Presbyterian church. At Princeton, Witherspoon, taught moral philosophy classes—as well as history and rhetoric—to revolutionaries such as James Madison. T. Miller explains that Witherspoon was interested not in the emerging belletristic tradition or epistemology rhetorics promoted by Hugh Blair and George Campbell, but rather "with the practical art of speaking to public controversies" (*Eighteenth-Century* 268). He delivered sermons supporting American Independence and was the only minister to sign the Declaration of Independence (*Eighteenth-Century* 268). According to Miller, Witherspoon, like other Scottish rhetoricians and philosophers before and after him (notably Hutcheson, Smith, and

George Jardine), "assumed that moral philosophy covered 'the whole business of active life,' of which the studies of rhetoric, composition, and criticism should be subordinate" (*Eighteenth-Century* 271). Under Witherspoon, however, and against the backdrop of revolutionary America, moral philosophy became instruction in public communication, ethics, and persuasion. More than his contemporaries, Witherspoon in many ways defined moral philosophy in rhetorical terms of Aristotle and Cicero, and viewed the theories of fellow Scotsmen Hutcheson and Smith in terms of rhetorical engagement and the rich tradition of civic humanism. For Witherspoon, the study of rhetoric and communicative practices came with ethical and political responsibilities. As in the classical period, rhetoric in Witherspoon's conception of moral philosophy became the catalyst for bridging theoretical classroom instruction and active community involvement. His blending of civic humanism with rhetorical instruction provides a model for moving instruction in professional and technical communication from isolated skills acquisition to social praxis, which includes an understanding of social implications and social responsibilities of the writer.

Obviously, the Scottish democratic ideal of education isn't the same "democratic ideal" we hold today, but the moral philosophy classes still have much to teach us as we revise the aims of writing instruction to include civic concerns. "If we truly feel that we writing teachers are the heirs of the rhetorical tradition," explains Bruce Herzberg, "we may take great comfort in making this connection the basis of a civic discourse pedagogy" (CD-ROM). Viewing the aims of professional communication in terms of civic rhetoric has the potential for changing the face of contemporary professional communication pedagogy.

APPLICATIONS FOR TECHNICAL AND PROFESSIONAL COMMUNICATION PROGRAMS

The eighteenth-century courses in moral philosophy and the civic rhetoricians who taught them offer valuable lessons for present-day professional and technical communication theories and practices. The Scottish educational system of the eighteenth century emphasized that universities and colleges prepare students for citizenship. Adam Smith, who set a precedent at the Universities of Edinburgh and Glasgow and, through his students, influenced American education, promoted educational philosophies dedicated to training citizens (Court, *Scottish Connection* 21). By the end of the eighteenth century, the University of Pennsylvania, founded on Scottish educational principles, included a Moral Philosophy Department that addressed the study of ethics, economics, politics, logic, and rhetoric (*Scottish Connection* 21, 24). These moral philosophy courses, with their intertwined and interdependent relationships among rhetoric, ethics, and writing, addressed students' self-interests with a rhetorical and ethical education, served the needs of the universities and cities, and prepared students to adopt the roles of community intellectuals.

Our courses in professional communication can achieve similar reciprocal relationships. Jim Porter proposes one way to view the relationship among rhetoric, ethics, and writing in light of civic humanism: "Ethics pertains to a process of inquiry by which we determine what is right, just, or desirable in any given cause. This sense of ethics requires rhetoric as a co-equal partner in the pursuit of a position—this is not ethics as a Final Truth, but as a standpoint, a contingent commitment" (29). Porter asserts that rhetoric and ethics cannot be separated and therefore "all writing has an aim,…is rhetorical, and therefore…all writing has an ethical component: It aims for the good of somebody or something" (68). He defines rhetorical ethics as "a set of implicit understandings between writer and audience about their relationship. Ethics in this sense is not an answer but is more a critical inquiry into how the writer determines what is good and desirable. Such inquiry necessarily leads toward a standpoint about what is good or desirable for a given situation" (68).

Any rhetorical situation or action calls for ethical inquiry because "rhetorical action always involves ethical judgment because the very act of composing establishes relations between writers and audiences, or relations depending on some notions of 'rightness' and 'wrongness'" (62). This relationship between communicators and their audience(s) is especially important to professional and technical communicators. All professional writing (business, technical, digital) addresses some situated action, which affects an audience. Professional and technical communication programs that emphasize rhetoric and ethical inquiry inherent in writing projects continue a tradition begun by moral philosophy professors of the eighteenth century who neither engaged in didactic teaching of right and wrong nor dictated to students what they should think in given situations.

The research addressing the various literacies associated with educating effective professional and technical communicators complements our discussion of civic rhetoric as well. Kelli Cargile Cook's article on literacies provides a theoretical framework for incorporating the teaching of a number of literacies within our programs and courses. Her identification of the multiple literacies and pedagogical examples of incorporating this multilayered approach to our programs and courses includes the following six literacies: basic, rhetorical, social, technological, ethical, and critical (7). Although current research addresses the multiple "literacies that technical communicators should acquire," which literacies and how to integrate and "layer" the learning of these literacies in our programs, courses, and assignments are the questions Cargile Cook addresses in her work.

These literacies can be layered through a civic-minded, rhetorical approach to our programs and courses. Of the six literacies, Cargile Cook's discussion of rhetorical, social, technological, ethical, and critical literacies provides effective descriptions of the literacies needed to form a foundation for the approach we are advocating. She identifies rhetorical literacy as the ability to understand the audience(s), purpose(s), context(s) associated with situations calling for rhetorical action, in conjunction with "an awareness of one's own ideological stance" (10). Social literacy refers to the ability to collaborate as well as being able to work within organizational settings. We must remember that professional and technical communicators who are socially literate may also be in positions to reform or change existing organizational settings—many of our students have the potential to become active agents of change. Historically, technological literacy meant primarily knowing how to use computer applications and/or writing technologies. However, proficiency in technology is only one component of technological literacy. Wahlstrom claims we must also provide students with opportunities "to explore the ways in which technological change affects community, authority, and ethics" (132). Ideally, professional and technical communicators should be prepared to serve as technology facilitators, information architects, and usability researchers (Cargile Cook 14). Both Cargile Cook and Wahlstrom argue for layering ethical literacy with other curricular goals to enhance the "technical communicator's abilities to make decisions that are grounded in the profession's ethical principles" and enhance decision making by "making them more cognizant of ethical implications of their decisions, including their responsibilities as citizens and workers in their society" (Cargile Cook 16). Cargile Cook defines critical literacy as "the ability to recognize and consider ideological stances and power structures and the willingness to take action to assist those in need" (16). These literacies are especially important to consider when arguing for a civic-minded and rhetorical approach to professional and technical communication courses and program design. They are also consistent with the goals of past moral philosophy courses: citizenship, professionalization, ethical action, and effective communication.

CREATING COMMUNITY INTELLECTUALS THROUGH CIVIC ENGAGEMENT

According to the *Stanford Encyclopedia of Philosophy*, the term "civic humanism" is most often associated with "positive connotations" of "public-spirited citizenship." Specifically, "civic humanism is linked in principle to a classical educational program that goes beyond the formative capacity of participatory citizenship itself and involves the conscious revival of ancient ideals" (Moulakis, pars. 3, 6). Adopting a combination of a civic-minded philosophy with service-learning pedagogy, defined as "a method by which students learn through active participation in thoughtfully organized service; is conducted in, and meets the needs of the community, is integrated into and enhances the academic curriculum; includes structured time for reflection and helps foster civic responsibility" (National Community Service Trust Act), in our courses can help ensure the education of community intellectuals who are interested in civic/community engagement.

T. Miller, arguing for the adoption of a civic stance in communication courses, reminds us that "a civic philosophy of rhetoric can enable us to bring our work with service-learning, new technologies, and political controversies into a unified project that challenges the hierarchy of research, teaching, and service that limits the social implications of academic work and devalues the work of the humanities" ("Rhetoric Within" 34). In providing educational settings that encourage students to develop a civic mindset, we help alleviate disciplinary tension between teaching practical skills and teaching rhetoric as a civic virtue. As a result of this approach, students will be prepared to apply their educational experiences to the communities in which they live, work, and serve.

It seems obvious that programs focusing on civic/community engagement might make service-learning, ethics, and civic rhetoric the foundations of their programs and thus their courses. We should be creating programs and courses that emphasize engagement of students with their communities, particularly in our service courses in business and technical writing. However, the notion of community does raise questions. We tend to think of community in terms of the civic community or university community, but businesses, corporations, and organizations are all communities in themselves, and they, in turn, serve many other communities. A study of communities through community engagement can fulfill students' other concerns as well. Civic rhetoric can allow students to link their own self-interests with public concerns. In order for civic rhetoric to make a difference in our programs and our courses, we must ensure that our service-learning projects are "connected to the day-to-day institutional work and ethos of the writing program and its faculty" (Grabill and Gaillet 73). Course projects selected for civic/community engagement should build on the criteria outlined by Grabill and Gaillet:

- The projects meet a real need as articulated by our community partner.
- The projects are sophisticated and writing related.
- The projects fit into the time frame given to the project. (72)

Students, universities, and communities can all benefit from community engagement. Students benefit in a wide variety of ways from participating in service-learning projects. They have the opportunity to apply their multiple literacies and learning within communities reflecting their own interests. Students are also able "to interact with real-world audiences" and receive feedback from their community partners as they engage in a symbiotic relationship. Students also learn to manage their writing projects and negotiate ethical dilemmas. As a result, they are able to establish relationships with community members and to create materials for a writing portfolio (Bowdon and Scott 13–17). Students continue to learn while participating in these projects, and they also apply what they already know. This seems obvious, but students who apply what they learn in a classroom to an outside context tend to have richer educational experiences and recognize how they might continue to learn once they leave their current course.

The organizations and the campus also benefit from service-learning projects by nurturing a relationship between the community and the college—often labeled a "town and gown" relationship. Businesses and organizations no longer assign a role to colleges and universities or dictate what students need to know and learn; colleges and universities no longer serve as merely disseminators of knowledge. A mutual relationship emerges where each party learns from the other and strives to meet the other's needs. This mutually beneficial "town and gown" relationship not only existed but also informed curriculum development and pedagogical practices in eighteenth-century Scottish universities, particularly at Glasgow and Edinburgh. T. Miller writes of the historical roots of service-learning, "ancient doctrines of public duties and divine missions often value learning by doing in the assumption that we can come to know what to do by putting what we know to good use. Contemporary discussions of service-learning stress the reciprocal dynamics of learning from others by sharing what we know how to do" ("Other Voices" 22).

The foundation of the relationship between technical and professional communication instruction and service-learning projects should include the principles of civic rhetoric and what Bowdon and Scott refer to as "rhetorical activism" (288). One way for community intellectuals to contribute to their communities is through acting rhetorically in specific situations for specific purposes. Professional and technical communication's "rhetoricity—that is, its situatedness and connection to specific audiences—makes service-learning particularly well suited to it. Service-learning projects are clearly situated forms of social, rhetorical action intended to help particular audiences solve particular problems" (30). Students who engage with local rhetorical situations are learning to become community intellectuals in much the same way that students took their place in various communities during the eighteenth century.

Professional and technical educators are often criticized for doing no more than educating students for their careers or teaching them skills so that they can get jobs. Pejorative terms used to describe the discipline of professional and technical communication include vocational, utilitarian, and practical (Dubinsky; C. Miller). However, approaching our programs with a combination of rhetoric, ethics, service-learning and moral philosophy provides an effective foundation for nurturing community intellectuals. Chris Anson explains, "Theories of service learning value reflection for helping to create the connection between academic coursework and the immediate social, political, and interpersonal experiences of community-based activities" (167). Bowdon and Scott also point out that "the practical nature of most technical and professional communication makes it easy to overlook the values associated with it and effects enabled by it. Yet, good service-learning work is more than rhetorically effective—it is also ethical and beneficial to the community" (6).

Service-learning and civic rhetoric go hand-in-glove because of the emphasis on rhetoric, ethics, and civic engagement. By adopting this blended pedagogy, we have the opportunity not only to educate our students for their chosen professions, but also to send them to community organizations and businesses equipped to question community constructions and engage in rhetorical practices. As a result of these experiences, we hope students see themselves as community intellectuals, who can use the skills and knowledge they learn in their own communities.

WORKS CITED

Anson, Chris M. "On Reflection: The Role of Logs and Journals in Service-Learning Courses." *Writing the Community: Concepts and Models for Service-Learning in Composition.* Ed. Linda Adler-Kassner, Robert Crooks, and Ann Watters. Washington, DC: American Association for Higher Education, 1997.

Bizzell, Patricia, and Bruce Herzberg. *The Rhetorical Tradition.* New York: Bedford/St. Martin's, 1990.

Bowdon, Melody, and J. Blake Scott. *Service-Learning in Technical and Professional Communication.* New York: Longman, 2003.

Cargile Cook, Kelli. "Layered Literacies: A Theoretical Frame for Technical Communication Pedagogy." *TCQ* 11.1 (2002): 5–29.

Court, Franklin E. *Institutionalizing English Literature: The Culture and Politics of Literary Study, 1750–1900.* Stanford: Stanford UP, 1992.

—. *The Scottish Connection: The Rise of English Literary Study in Early America.* Syracuse, NY: Syracuse UP, 2001.

Cushman, Ellen. "The Public Intellectual, Service-Learning, and Activist Research." *CE* 61.3 (1999): 328–36.

—. "The Rhetorician as an Agent of Social Change." *CCC* 47 (1996): 7–28.

Davie, George. *The Democratic Intellect.* Edinburgh: Edinburgh UP, 1982.

Deans, Thomas. *Writing Partnerships: Service-Learning in Composition.* Urbana, IL: NCTE, 2000.

Dubinsky, James M. "Service-Learning as a Path to Virtue: The Ideal Orator in Professional Communication." *Michigan Journal of Community Service Learning* 8.2 (2002): 61–74.

Enos, Richard Leo. "Cicero." *Encyclopedia of Rhetoric and Composition.* Ed. Theresa Enos. New York: Garland, 1996. 102–05.

Farmer, Frank. "Review: Community Intellectuals." *CE* 65 (2002): 202–10.

Findlay, Ian. *Education in Scotland.* Hamden, CT: Archon Books, 1973.

Grabill, Jeffrey T., and Lynée Lewis Gaillet. "Writing Program Design in the Metropolitan University: Toward Constructing Community Partnerships." *Writing Program Administration* 25.3 (2002): 61–68.

Herzberg, Bruce. "Civic Literacy and Service Learning." *Coming of Age: The Advanced Writing Curriculum.* CD-ROM insert. Ed. Linda K. Shamoon, Rebecca Moore Howard, Sandra Jamieson, and Robert A. Schwegler. Portsmouth, NH: Boynton/Cook, 2000.

Horner, Winifred Bryan. *Nineteenth-Century Scottish Rhetoric: The American Connection.* Carbondale: Southern Illinois UP, 1993.

—. "Nineteenth-Century Rhetoric at the University of Glasgow with an Annotated Bibliography of Archival Materials." *RSQ* 19 (1989): 173–85.

Howell, Wilbur Samuel. *Eighteenth-Century British Logic and Rhetoric.* Princeton: Princeton UP, 1971.

Huckin, Thomas. "Technical Writing and Community Service." *JBTC* 11.1 (1997): 49–59.

Hunter, Leslie S. *The Scottish Educational System.* Oxford: Pergamon, 1966.

Miller, Carolyn. "A Humanistic Rationale for Technical Writing." *CE* 40.6 (1979): 610–17.

—. "What's Practical About Technical Writing?" *Technical Writing: Theory and Practice.* Ed. Bertie E. Fearing and W. Keats Sparrow. New York: MLA, 1989. 14–24.

Miller, Thomas. *The Formation of College English: Rhetoric and Belles Lettres in the British Cultural Provinces.* Pittsburgh: U of Pittsburgh P, 1997.

—. "John Witherspoon." *Encyclopedia of Rhetoric and Composition.* Ed. Theresa Enos. New York: Garland, 1996. 767–68.

—. "John Witherspoon." *Eighteenth-Century British and American Rhetoric and Rhetoricians: Critical Studies and Sources.* Ed. Michael G. Moran. Westport: Greenwood, 1994. 268–80.

—. "Other Voices 2.B." *Service-Learning in Technical and Professional Communication.* Ed. Melody Bowdon and J. Blake Scott. New York: Longman, 2003. 22.

—. "Rhetoric Within and Without Composition: Reimagining the Civic." *Coming of Age: The Advanced Writing Curriculum.* Ed. Linda K Shamoon, Rebecca Moore Howard, Sandra Jamieson, and Robert A. Schwegler. Portsmouth: Boynton/Cook Heinemann, 2000. 32–41.

—. "Treating Professional Writing as Social *Praxis.*" *JAC* 11.1 (1991): 57–72.

Moulakis, Athanasios. "Civic Humanism." 1 Oct. 2003. *Stanford Encyclopedia of Philosophy.* 35 pars. Ed. Edward N. Zalta. Winter 2002 ed. Metaphysics Research Lab. Stanford University. 2002. <http://plato.stanford.edu/entries /humanism-civic/>.

National Community Service Trust Act of 1993, Pub. L 103-83. 21 Sept. 1993. Stat. 107. 785–923.

Porter, James. *Rhetorical Ethics and Internetworked Writing.* Greenwich, CT: Ablex, 1998.

Posner, Richard. *Public Intellectuals: A Study of Decline.* Cambridge: Harvard UP, 2002.

Rubens, Philip. "Technical and Scientific Writing and the Humanities." *Research in Technical Communication.* Ed. Michael Moran and Debra Journet. Westport: Greenwood, 1985. 3–23.

Smith, Adam. *Lectures on Rhetoric and Belles Lettres.* Ed. J.C. Bryce. New York: Oxford UP, 1983.

Ulman, Lewis. "Adam Smith." *Eighteenth-Century British and American Rhetoric and Rhetoricians: Critical Studies and Sources.* Ed. Michael G. Moran. Westport: Greenwood, 1994. 207–18.

—. "Adam Smith." *Encyclopedia of Rhetoric and Composition.* Ed. Theresa Enos. New York: Garland, 1996. 674–75.

Wahlstrom, Billie J. "Teaching and Learning Communities: Locating Literacy, Agency, and Authority in a Digital Domain." *Computers and Technical Communication: Pedagogical and Programmatic Perspectives.* Ed. Stuart Selber. Greenwich, CT: Ablex, 1997. 129–46.

Michelle F. Eble is Assistant Professor of Technical and Professional Communication in the Department of English at East Carolina University, where she teaches courses in professional writing, publications development, history of professional communication, ethical issues in professional communication, and editing. Her work on writing technologies and the teaching of professional writing has appeared in *Computers and Composition* and *Technical Communication.*

Lynée Lewis Gaillet is Associate Professor of Rhetoric and Composition at Georgia State University, where she serves as Director of Lower Division Studies. She is the editor of *Scottish Rhetoric and Its Influences,* and author of numerous articles and book chapters examining the history of modern writing practices. Her work has appeared in journals such as *Rhetoric Society Quarterly, Journal of Advanced Composition, Journal of Basic Writing, Rhetoric Review, Writing Program Administrator,* and *Composition Studies.*

TECHNICAL COMMUNICATION QUARTERLY, *13*(3), 355–368

REVIEWS

Tracy Bridgeford, University of Nebraska at Omaha, Editor

Gerard A. Hauser and Amy Grim. *Rhetorical Democracy: Discursive Practices of Civic Engagement.* **Mahwah, NJ: Lawrence Erlbaum Associates, 2004. 321 pp.**

Reviewed by Evie Johnson
Michigan Technological University

Rhetorical Democracy: Discursive Practices of Civic Engagement offers students, scholars, and teachers a cache of rhetorical analysis and discussion topics. As I read the articles (proceedings from the 2002 Rhetoric Society Conference in Las Vegas), I was preparing to cover issues of democracy and education with my class of English Education preservice teachers. This collection both resonated with and complicated my plans. Where my original teaching plan (to analyze classroom cases across America) bordered on a celebration of education as civic engagement, *Rhetorical Democracy* challenged me to attend to the rhetorical issues of teaching, to "tend to the messy business of a democracy's rhetoric. That project requires reconsidering democratic life in light of a number of factors that are not part of the Athenian legacy on which so much of our theory and criticism of rhetoric rests" (3). This book brings to the surface a number of those factors (e.g., cyberculture, terrorism, racism, and so on), laminates them, and pushes me to think differently about civic engagement and citizen roles.

Not only does *Rhetorical Democracy* challenge my thoughts on teacher training pedagogy as a rhetorical act, it pressures me to reconsider my teaching of technical communication and civic engagement. Indeed, other technical communication teachers who take a rhetorical approach, as well as students in introductory rhetoric classes at the graduate or undergraduate level, are just two of the many audiences that will find this collection useful, if not central, to their classroom discussions.

I see especially strong potential for using this text in graduate rhetoric classes. Its thirty-nine chapters, distributed among three sections, are short—around seven pages—and accessible, introducing students to rhetoric through contemporary as well as historical examples. Students would become familiar with the roots of rhetorical studies on democracy and be motivated to seek out the primary texts repeatedly referenced. Using such a book to open a rhetoric class honors the pedagogical principle of grounding new knowledge in students' experiences and presenting new information that draws that knowledge toward new understandings. One can't help but reflect on one's own civic engagements in reading this book. Whether the reader has built a Web page or attempted to challenge a local ordinance at a town council meeting (or knows someone who has), this book provides a chapter that will resonate and locate that experience historically and rhetorically as a type of civic engagement. Let me sample just a few of the opportunities this book offers readers.

John Killoran's chapter, "Homepages, Blogs, and the Chronotopic Dimensions of Personal Civic (Dis-)Engagement," explores genre theory and the analysis of Web-based personal publications. He explores Bakhtin's concept of chronotope ("time-space" background) comparing familiar Web genres no doubt popular with many readers, particularly students. He also draws from Carolyn Miller's 1984 article "Genre as Social Action" to articulate the potential for and type of social action made available by these two genres. Killoran claims that because of the chronotopes of these different genres, they operate differently as discursive tools of engagement. Specifically, because of the chronotopic positioning of each genre, the homepage is more conservative and the blog is more radical: "In the new media environment favoring currency and enabling easy contact among citizen publishers, it is, of these two genres, the blog that may better sustain a more distinctive, more socially responsive means of civic engagement " (218). Surely readers will have reactions worthy of debating such a claim.

Readers will also find "The Rehabilitation of Propaganda: Post 9/11 Media Coverage in the United States" by Mark Andrejevic a thought-provoking and cutting-edge analysis of post-9/11 mainstream media, foreign policy, and the role of the public. Andrejevic asserts,

> Those who believe that any policy can be marketed as democratic if the PR department is good enough clearly don't believe that democratic principles have any determinate content or any nonarbitrary relation to material reality. They seek not to confront contradictions but to paper them over, and in so doing posit an incoherently one-sided relationship between discourse and material reality: Determination can flow from the former to the latter, but not the other way around. (90)

Material reality is one of the messy areas technical writers are trained to clean up. Unhinging (or lubricating) the sometimes uneasy articulations of communication and material reality is familiar terrain to technical communication—but is civic engagement? Clearly, the attempt is made in technical communication courses to impress considerations of audience upon students as a way to clear the path between audience and material reality. But we don't always bring concepts of democracy to the drawing board. Yet, students of technical communication must learn to struggle with the contradictions inherent in all communication, their responsibilities as language mediators, their identities as often "silent" authors, and their participation in both their future workplaces and civic forums. I would use

Andrejevic's chapter and others contained in this section to examine issues of social responsibility, communication, and the workplace—just to expose the messi-
ness even more.

One of the best ways to examine social responsibility and civic engagement on
campus or off is described by Herbert Simons, in one of the four plenary papers.
His chapter, "The Temple Issues Forum (TIF): Innovations in Pedagogy for Civic
Engagement," occupies (and takes) an important threshold position in the collection and is among a group of chapters that prepare readers to engage the issues sub-
sequently presented in the selected papers; it also invites university faculty to organize formal gatherings where community members can practice civic engagement
with local and global issues.

The plenary papers section, however, does not merely draw readers toward the
democratic possibilities of rhetorical democracy. It draws readers into the problems
and the arguments that follow in the book (and in the public forum as well). For example, although Simons's discussion of the Temple Issues Forum cautiously recom-
mends university-organized civic engagement, Rolf Norgaard's "Desire and Performance at the Classroom Door: Discursive Laminations of Academic and Civic
Engagement" inspects Simons's themes under a slightly different university light:
that of the composition classroom. He applies the concept of "laminations" to explore how our academic and civic roles constitute an admixture. Norgaard wants us
to examine the presumptions of compositionists' "civic fantasy." Regarding some
compositionists' desire to link text and street, he warns, "We must not congratulate
ourselves too quickly when we turn the traditional academic essay into an op-ed
piece, proposal, or a letter to an elected official. Instead, we and our students might
reflect on the paradoxes of that desire" (257). This chapter tugs at the ligaments of
contemporary composition instruction's theories and pedagogies in ways that will
have readers probing the purposes and methods of college writing instruction.

Ultimately, this collection is like a vibrant community: it speaks from different
viewpoints, asks important questions, and offers many directions for those whose
goals include communicating across the inevitable divides of civic life. And like a vibrant community, it takes readers across a varied terrain of worries and joys. Gerald
Hauser's introduction names the biggest source of joy and worry for his readers—developing a "sense of how to productively encounter others in a community of
strangers…[c]ommunicating across the divide is perhaps the central issue confront-
ing the current discussion about civic engagement" (13). He encourages readers,
most of whom will be located in the academy, to "focus our instruction in rhetoric as
the place to develop the links of understanding and common cause among differences. These are the relations on which democratic culture depends" (13). Although
we can dream of a national curriculum that emphasizes civic engagement, like the
one that propelled science and math curricula during the Cold War, current trends
don't seem to be moving there quite yet—more reason to take up the call.

Although it may seem formidable, the call of rhetorical democracy is actually a
task we take up each day through civic engagment in our classrooms, homes, and

workplaces. Hauser's suggestion—"to think of it as enabling our students to live as free human beings who have it within their power to influence the communities in which they will work, make their homes, form friendships, raise families, educate their children, enjoy public arts, and pursue their private pleasures" (13)—implores us to reframe a national curriculum for civic engagement as one that is immanent, community-based, and close at hand.

Putnam, Robert D. "Bowling Alone: America's Declining Social Capital." *Journal of Democracy* 6.1 (1995): 65–78.

Spiezio, K. Edward. "Pedagogy and Political (Dis)Engagement." *Liberal Education* 88.4 (2002): 14–19.

Reviewed by Amy C. Kimme Hea
University of Arizona

In "Bowling Alone: America's Declining Social Capital," Robert D. Putnam, the noted Dillon Professor of International Affairs and Director of the Center for International Affairs at Harvard University, proposes "social capital" as key to an active citizenry. In "Pedagogy and Political (Dis)Engagement," K. Edward Spiezio, a respected associate professor of politics and the Executive Director of the Participating in Democracy Project at Cedar Crest College, suggests that in this post–September 11 climate, we educators can foster long-term political engagement through innovative curricular change. Despite their disciplinary differences, Putnam and Spiezio argue that American society can benefit from more active civic and political participation. As professional and technical communication scholars and teachers, we too have a stake in better understanding and fostering both civic and political engagement. In this review, I outline Putnam's view on social capital and Spiezio's research on political curricula and discuss the implications of their research for those of us in professional and technical communication, especially those of us teaching service-learning projects.

"BOWLING ALONE: AMERICA'S DECLINING SOCIAL CAPITAL"

Drawing on social science theory, Putnam defines social capital as "features of social organization such as networks, norms, and social trust that facilitate coordination and cooperation for mutual benefit" (67). Current data argue that strong social communities contribute to the development of civic engagement, economic growth, and performance of representative government. Concerned about eroding political participation in the United States, Putnam holds out hope that with a better understanding of social capital, social scientists can create strategies for re-vitalizing the body politic.

Putman turns his discussion of social capital toward quantitative analyses of American political disengagement over the past four decades. Citing statistics concerning the decline in voter turnouts, pervasive mistrust of Washington government, and reduction of public participation in town and school meetings, political speeches, and rallies, Putnam argues that such a climate is attributable not merely to "strictly political explanations" but also waning participation in personal, face-to-face, community-based organizations. Interpreting the aggregate results from the fourteen-time, national General Social Survey, Putnam notes the "modest decline" of church going and church-related service organizations despite the fact that "the United States has more houses of worship per capita than any other nation on Earth" (69). Further, Putnam argues that organizations which typically drew large numbers of women supporters, such as school services, sports groups, professional organizations, and literary societies, have all witnessed a loss in numbers. Moreover, Putnam contends that traditionally male groups like sports clubs, labor unions, professional societies, fraternal groups, veterans' groups, and service clubs have experienced loss of membership throughout the past decades. Putnam laments the loss of civic participation in organizations such as labor unions, the Parent Teacher Association (PTA), Federation of Women Voters (LVW), Federation of Women's Clubs, Boy Scouts of America, Lions, and Elks. He, however, turns to the sport identified in his title for his most "whimsical yet discomforting bit of evidence of social disengagement in contemporary America" (70). He notes that, despite a forty percent increase in the popularity of bowling from 1980 to 1993, bowling leagues—the social capital aspect of the sport—have declined by ten percent over the same time period.

To complicate his analysis of social capital as defined by organizational participation, Putnam cites the "countertrends" of developing new organizations, nonprofits, and support groups (70). Looking at the growth of "vibrant new organizations," or what he later terms "tertiary associations," Putnam identifies rising memberships in environmental groups, feminist organizations, and the American Association of Retired Persons (AARP) (71). Putnam concedes the political influence of these organizations but denies their "social connectedness," especially as compared with other previously named associations (71). To make his point, Putnam asserts that there is little bond among members because those persons likely never meet or exchange views. Putnam stipulates that one condition of social capital is that membership increases social trust, a goal that cannot be guaranteed by such tertiary affiliations. Likewise, Putnam explores the increase in nonprofit membership. Pointing to the sheer number and range of nonprofit groups from "Oxfam and the Metropolitan Museum of Art to the Ford Foundation and the Mayo Clinic" and their far-ranging missions, he asserts that nonprofits do not foster social connectedness (71). Finally, turning to the rise in support group participation, Putnam assents to their usefulness in fostering social trust and connectedness, but he and Robert Wuthnow, whose work he cites, argue that "small groups may not be fostering community as effectively as many of their proponents would like" (qtd. in Wuthnow 3–6). Much of Putnam's point is that these forms of social interaction do not meet the conditions of social capital—social trust, connectedness, and collectivity—and thus cannot discredit his claims that organizational membership can be correlated with civic engagement.

Putnam continues his exploration with an assessment of potential causes of civic disengagement: (1) "the movement of women into the labor force"; (2) "mobility: the 're-potting' hypothesis"; (3) "other demographic transformation"; and (4) "the technological transformation of leisure" (74–75). Regarding the move of women into the workforce, Putnam states that "it seems highly plausible that this social revolution [of women in the workforce] should have reduced the time and energy available for building social capital" (74). Likewise, men, who have taken up some of the household responsibilities, also have reduced memberships. His conclusion, however, is that another factor "besides the women's revolution seems to lie behind the erosion of social capital" (74). Considering the factor of social mobility, Putnam reports that despite increased homeownership, civic participation is still on the decline. Other demographic concerns for Putnam include "fewer marriages, more divorces, fewer children, [and] lower real wages," all of which, he remarks, could be impacting civic engagement, "since married, middle-class parents are generally more socially involved than other people" (75). He associates these changes with larger economic phenomenon such as decreased local markets and increased mega and online stores. For Putnam, the eroding of social capital also is influenced by technology; he claims that "the most powerful instrument of revolution is television...[which] has made our communities (or, rather, what we experience as our communities) wider and shallower" (75). Technology could be dividing our individual and collective interests in Putnam's opinion, and it is worthy of further study.

Putnam closes by suggesting that social scientists pursue lines of inquiry that identify which types of organizations and networks most effectively represent and create social capital; empirically assess the role of technology and the workplace on the development of social capital; analyze the benefits and costs of community engagement; and establish connections between public policy and social capital. Putnam seeks an increase in social capital through membership in traditional organizations and face-to-face exchange. To this end, Putnam asserts "high on America's agenda should be the question of how to reverse...adverse trends in social connectedness, thus restoring civic engagement and civic trust" (77).

PEDAGOGY AND POLITICAL (DIS)ENGAGEMENT

Citing the September 11th tragedy as a catalyst to younger Americans' civic and political participation, Spiezio speculates on whether the renewed dedication to political engagement is just a "temporary phenomenon" or a "fundamental change" (14). Hoping for the latter, Spiezio argues that educators must seize this historical moment and create pedagogies that engender stronger social and political activism. Based upon his research, Spiezio explains that to promote political participation, teachers and administrators must deploy explicitly political learning strategies. His article articulates some potential approaches to improve educational practices in service of active citizenship.

Through the three-year, $1.2 million Participating in Democracy Project at Cedar Crest College, Spiezio and his colleagues Elizabeth Meade and Suzanne Weaver have found that current educational curricula either thwart or provide limited engagement in the development of political and civic consciousness. Further, their research has discovered that students are not gaining the knowledge, skills, and confidence to engage in political processes. To support these claims, Spiezio analyzes the results of the annual, nationwide telephone survey, Campus Attitudes toward Politics and Public Service Survey (CAPPS). Conducted since 2000 by the Institute of Politics at Harvard University and developed by and administered to undergraduates, the survey asks college students from around the nation to rank practices that will help "promote a greater degree of political engagement among younger Americans" (15). According to data from 2000 and 2001, 80 percent of students responded that the following actions would be somewhat or very effective in this mission:

- more time teaching the basics about how to get involved in politics, activism, and issues of the day;
- easy-to-find Web sites dedicated to providing students with political information;
- college partnerships with local and state government that offer academic credit to students who participate in public activities;
- direct student contact with more elected officials, members of government, political candidates, campaigns, and institutions;
- student-oriented political action committees or networks that focus on organizing student groups, training students for political involvement and helping young people get elected to local, state, and federal offices;
- increased awareness of real-life examples of how young people can make a difference politically; and
- making registering and voting by absentee ballot easier for college students. (15–16)

Spiezio notes that "85 percent of undergraduates felt that 'community volunteering' was better than 'political engagement' as a way of solving community problems" and that "an overwhelming majority of students believe that 'volunteering in the community is easier than volunteering in politics'" (15). As Spiezio later reveals, an emphasis on civic versus political engagement is troubling for him and other political scientists.

Rather than arguing that student responses represent a cynicism about politics, however, Spiezio speculates that students feel less confident in their knowledge of politics, evidenced by their 85 percent agreement with the statement "I feel like I need more practical information about politics before I get involved" (16). For Spiezio, this lack of knowledge and confidence in politics can be traced to two important factors: an emphasis on "apolitical" or nonexplicitly political models of service-learning and an approach to civic education that promotes passive democratic participation.

Taking up the first factor, Spiezio maintains that, because a "communitarian" approach to service-learning curricula has grown over the past decade, students are not asked to think beyond their own individual contributions to and conceptions of effective community action (16). For researchers like Harry C. Boyte and Nancy N. Kari, whose research Spiezio draws upon, "communitarian versions of citizenship tend to separate ideals like community, the common good, and deliberation from the messy, everyday process of getting things done in a public world of diverse interests" (41). Boyte and Kari affirm Mary Hepburn's review of service-learning projects that posits to "build attitudes of political efficacy and civic involvement, the service and related curriculum content should include government, political issues, and/or social action" (qtd. in Hepburn 48–49). Spiezio raises this problem of service-learning not to dispute its pedagogical effectiveness but to sharpen its political focus. He asserts that for service-learning to affect student political consciousness and practices, this method must explicitly and directly integrate politics, not just ethics.

Disturbed by the lack of political engagement on the part of America's youth, Spiezio finds little comfort in civics curricula that attend to an individual's needs and moral life rather than the larger, complex political community. Terming it "political passivity," Spiezio is disheartened by civics pedagogies that rely on representative democracy and consumer-based, writing-your-congressperson approaches to government involvement (17). Spiezio sees such teaching as exacerbating the problems of political disengagement by creating passive citizens whose actions are guided through already accepted forms of state-centered practice, of which voting is the primary example. Thus, students are not learning to create policy, affect change, or become political activists.

Spiezio redresses the curricular shortcomings of both service-learning and civic pedagogies suggesting that educators can encourage political engagement by teaching students to

- learn the basic strategies and tactics of political activism, complemented perhaps with an understanding of the fundamentals of applied policy analysis;
- interact with practitioners who have chosen public administration, or politically meaningful forms of public service more generally, as a vocation;
- network with student-based organizations and citizen-based interest groups that are politically active in various issue-areas; and

- experience political processes directly through placements in community-based institutions and organization that deal with applied policy issues. (18)

Acknowledging that faculty and administrative commitments are necessary to achieve these aims, Spiezio offers the success and dedication of the "Democratic Academy" as one model (19). As part of the Participating in Democracy Project, this organizational framework facilitates a partnership among Cedar Crest College, Heidelberg College, Lesley University, and St. Thomas Aquinas College. This collaborative project is an "integrated educational strategy designed to broaden and deepen the significance students attach to the meaning of citizenship and participation democracy through the creation of learning environments explicitly devoted to the promotion of political participation as a distinctive type of civic practice" (19). It is this project and others like it that will allow educators to channel the political and civic energies surrounding the September 11th tragedy to achieve long-term commitment to political engagement.

IMPLICATIONS OF THESE PROJECTS FOR PROFESSIONAL AND TECHNICAL COMMUNICATION

Both Putnam and Spiezio raise significant issues for those of us teaching service-learning projects in our professional and technical communication courses. The most important issue is the cultural trend of focusing on the individual as potentially undermining collective civic and political action. Many teachers in professional and technical communication have long-standing collaborative practices that attempt to complicate notions of individual authorship and encourage student critical thinking. These practices often extend to our service-learning pedagogies through team projects or other forms of group interaction. The researchers' appeals for further study to develop civic and political strategies are also equally significant to our interests. In fact, our field supports reflective research and teaching practices, evidenced by our publications, professional conferences, and active listservs. Based upon these broader issues, the articles serve as a touch point to continued discussion on how best to engage students in civic and political action through service-learning curricula.

In relation to these shared issues, however, these articles also make distinct arguments. In his work, Putnam creates a compelling set of factors for social capital—social trust, connectedness, and collectivity—but he then applies his definition in often limited and exclusionary ways. It would be difficult, for example, for a professional and technical communication instructor teaching service-learning projects not to notice that Putnam's definition of social capital excludes student participation in service-learning projects with nonprofit groups. This exclusion is based on his sense that (1) many nonprofit groups are tertiary and thus not dedicated to civic engagement, and (2) students cannot claim social trust or connectedness with groups unless they are official members. Putnam further limits his usefulness for many of us in the field by applying uncomplicated demographic categories (for example sex rather than gender) and determinist views of technology (claiming that technologies inevitably lead to social isolation). Although Spiezio does support Putnam's claim that nonprofits cannot constitute a site for political participation, he does stipulate that service-learning models must actively address politics in order to influence student political engagement. Making his argument less about particular organizations and more about reinvigorating curricula to restore political involvement, Spiezio offers specific curricular advice to teachers and administrators, the type of advice that many in our field already apply and that others can seriously consider. Finally, although I do not want to disparage the importance of quantitative research, I believe that both Putnam and Spiezio could answer their respective research calls by applying qualitative practices to enrich their projects. With continued study, a broader range of methodological inquiries, and even more complicated categories of analyses, Putnam, Spiezio, and other scholars working across disciplinary boundaries can help us in professional and technical communication better foster civic and political engagement through our research and teaching.

WORKS CITED

Boyt, Henry C., and Nancy K. Kari. "Renewing the Democratic Spirit in American Colleges and Universities: Higher Education and Public Work." *Civic Responsibiltiy and Higher Education.* Ed. Thomas Herlich. Phoenix: Onyx Press, 2000. 41.

Hepburn, Mary. "Service Learning and Civic Education in the Schools: What Does Recent Research Tell Us?" *Education for Civic Engagement in Democracy: Service Learning and Other Pormising Practices.* Ed. Sheilah Mann and John J. Patrick. Bloomington: Educational Resources Information Center, 2000. 48–49.

Wuthnow, Robert. *Sharing the Journey: Support Groups and America's New Quest for Community.* New York: Free Press, 1994.

Bowdon, Melody and J. Blake Scott. *Service-Learning in Technical and Professional Communication.* **New York: Addison Wesley Longman, 2003. 416 pp.**

Reviewed by Nora Bacon
University of Nebraska at Omaha

Ten years ago, when writing teachers talked about service-learning in composition or technical communication courses, we began with definitions of service-learning and often ended with a simple recommendation: "This is a good idea! Try it!" In a striking sign of the field's growth, today we can assume that writing teachers are familiar with the concept of service-learning—and in a sign of its maturation, we now distinguish among several approaches to service-learning and are prepared to consider which approach is most appropriate to one writing course or another. In *Writing Partnerships: Service-Learning in Composition*, Thomas Deans offers a useful taxonomy of service-learning approaches. Students can, he suggests, write about the community (doing volunteer work at community sites and then writing essays reflecting on their experience), write with the community (collaborating on poems, oral histories, position statements for community forums), or write for the community (producing brochures, manuals, newsletters, fundraising materials, and other documents for nonprofit organizations). Although any of these approaches can be adapted in a surprising variety of ways, I expect that each will finally settle into its most comfortable institutional home, with writing-about projects housed primarily in first-year composition courses, writing-with projects being assigned to more advanced students, and writing-for projects becoming a staple of courses in professional and technical communication.

Melody Bowdon and J. Blake Scott's excellent new textbook, *Service-Learning in Technical and Professional Communication*, demonstrates the match between the objectives of technical communication courses and the potential of assignments that involve writing for the community. The book is designed to support student writers through a semester-long, community-based writing project. Like writing internships, service-learning projects present students with real rhetorical problems, situated in real rhetorical contexts. Unlike internships, service-learning courses keep one foot firmly in the classroom, where the teacher and classmates are available to offer guidance, feedback, and—if the class is following Bowdon and Scott's model—a series of supplementary writing assignments that encourage close, thoughtful examination of every step in the community-based project.

The hallmark of the book is integration. It is true that the opening chapters seem a bit disjointed. Chapter 3, with its "toolbox" of rhetorical concepts, skips quickly from discourse-community theory to Aristotle's appeals to a lengthy discussion about the five canons that squeezes in everything from sequential and categorical organization plans to serif and sans serif fonts. But all these concepts return in subsequent chapters, where they are explained in more detail, and their relevance becomes clear.

In the heart of the book, Chapters 4 through 10, Bowdon and Scott offer a sequence of assignments designed to advance students' work on a collaborative community-based writing project, including

- a letter of inquiry addressed to a community agency that interests the student, introducing the project and soliciting a partnership;
- a resume to accompany the letter of inquiry;

- trip reports summarizing the results of preliminary conversations with the community contact person;
- a proposal, addressed to both the contact person and the instructor, articulating the project's goals and the writing team's plans;
- a field journal recording meetings and other events;
- a style sheet;
- a "discourse analysis memo" describing the agency and its characteristic discourse conventions;
- a progress report;
- a user-test report;
- an evaluation report;
- a presentation to the class and/or the community; and
- a transmittal letter to accompany the final draft of the community-based document(s).

Of course, students working on an ambitious project for the community—especially lengthy, complex documents like those displayed in the book's appendix—would not have time to complete all these supplementary assignments. But some of them (e.g., the proposal, the "discourse analysis," the progress report) are crucial, and instructors will be grateful for the wide array of choices.

As each assignment is presented, Bowdon and Scott discuss its audience and purpose, emphasizing the genuine rhetoricity of these assignments as well as the community-based document. They offer an example written by students from their classes and walk through it, discussing the genre in terms of its purpose and its structure. In several chapters, they also integrate lessons on style—using the letter of inquiry assignment, for example, as an occasion to emphasize the importance of concision and to show how fatty prose can be trimmed. Finally, and perhaps most important, they include sound, sensible advice for negotiating the social interactions that surround a real-world writing task, encouraging a constructive, problem-solving approach to the difficulties of collaboration on a writing team and the particular challenges of maintaining a campus-community partnership. Their advice is supplemented with a wealth of examples that could only have come from years of experience with many, many student writers on many service-learning projects.

What readers will find in Bowdon and Scott's book, then, is a wonderfully usable book, one that can guide a novice service-learning instructor though a successful course and provide experienced instructors with good, innovative ideas and food for thought.

What readers will not find is insight into the political implications of community-based work. Bowdon and Scott's course is resolutely preprofessional: Their primary concern is students' development as technical communicators, and they explicitly discourage involvement with the community organization that extends beyond the negotiated writing project. Although this makes sense in terms of project management, it is nevertheless problematic. Students who limit their role to "writing consultant" are likely to have extended contact with staff members at community organizations (usually middle-class professionals) but minimal contact with the organizations' clients. Their service-learning experience, then, introduces them to a narrow slice of the "community" so that, in the case of students whose backgrounds are middle-class or privileged, a valuable learning opportunity goes untapped.

Service-Learning in Technical and Professional Communication is one of three textbooks that Longman has published this year for use in service-learning writing courses. Consider this observation from Ross and Thomas's *Writing for Real*.

> The term "contact zone"…gives us an important framework in which to understand service-learning. …[S]ince most nonprofit organizations exist in order to fill a perceived void in information or services or in opposition to mainstream policies or practices, the nonprofit agency with which you work in your service-learning placement, as well as the constituency it serves, is likely to be marginalized in some respect. The organization's engagement with the dominant culture creates a powerful and dynamic contact zone, one in which change is possible. (34–35)

Ross and Thomas invite students to consider the "the power equation in 'service:'" the notion of *noblesse oblige*, the need for respectful reciprocity, the power attendant not only on racial majority status, economic privilege, and physical ability, but on education itself (42–53).

This sort of inquiry is missing from Bowdon and Blake. My hunch is that the vocational orientation of their course—their explicit (and probably highly successful) effort to prepare their students to land professional jobs and to function effectively in workplaces—is responsible for this omission. Bowdon and Blake encourage a critical perspective about community-based work and present it as an ethical matter, a question of what to do if a service-learning project brings to light unethical practices at a community agency or how to reconcile conflicts between the agency's values and the student writer's personal values. But if the project raises larger, political questions—about the social structures that create a need for a social service, about whose interests are served by the structures, about the relationship between service and activism—students or teachers wishing to reflect on the questions will be on their own.

If my students could take only one service-learning course, I might prefer that it be one where their range of contacts in the community and the range of critical questions raised about the experience were broader. But I have no wish to minimize Bowdon and Scott's accomplishment. Their textbook and the course whose scaffolding it provides are first-rate. If my students could take only one course in technical and professional communication, I would be delighted if it were this one.

WORKS CITED

Deans, Thomas. *Writing Partnerships: Service-Learning in Composition*. Urbana: NCTE, 2000.
Ross, Carolyn, and Ardel Thomas. *Writing for Real: A Handbook for Writers in Community Service*. New York: Addison-Wesley, 2003.

2004 SUBSCRIPTION ORDER FORM

Please ❏ enter ❏ renew my subscription to:

TECHNICAL COMMUNICATION QUARTERLY

Volume 13, 2004, Quarterly — ISSN 1057–2252/Online ISSN 1542–7625

SUBSCRIPTION PRICES PER VOLUME:

Category:	Access Type:	Price: US-Canada/All Other Countries
❏ Individual	Online & Print	$50.00/$80.00

Subscriptions are entered on a calendar-year basis only and must be paid in advance in U.S. currency—check, credit card, or money order. Prices for subscriptions include postage and handling. **Journal prices expire 12/31/04.** NOTE: Institutions must pay institutional rates. Individual subscription orders are welcome if prepaid by credit card or personal check. **Please note:** A $20.00 penalty will be charged against customers providing checks that must be returned for payment. This assessment will be made only in instances when problems in collecting funds are directly attributable to customer error.

❏   **Check Enclosed** (U.S. Currency Only)   **Total Amount Enclosed $**_____

❏   **Charge My:** ❏ VISA   ❏ MasterCard   ❏ AMEX   ❏ Discover

Card Number _____   Exp. Date____/____

Signature_____

(Credit card orders cannot be processed without your signature.)
PRINT CLEARLY for proper delivery. STREET ADDRESS/SUITE/ROOM # REQUIRED FOR DELIVERY.

Name_____

Address_____

City/State/Zip+4 _____

Daytime Phone #_____E-mail address_____
Prices are subject to change without notice.

For information about online subscriptions, visit our website at www.LEAonline.com

Mail orders to: **Lawrence Erlbaum Associates, Inc.,** Journal Subscription Department
10 Industrial Avenue, Mahwah, NJ 07430; **(201) 258–2200;** FAX (201) 760–3735; journals@erlbaum.com

LIBRARY RECOMMENDATION FORM

Detach and forward to your librarian.

❏ I have reviewed the description of *Technical Communication Quarterly* and would like to recommend it for acquisition.

TECHNICAL COMMUNICATION QUARTERLY

Volume 13, 2004, Quarterly — ISSN 1057–2252/Online ISSN 1542–7625

Category:	Access Type:	Price: US-Canada/All Other Countries
❏ Institutional	Online & Print	$140.00/$170.00
❏ Institutional	Online Only	$115.00/$115.00
❏ Institutional	Print Only	$125.00/$155.00

Name_____Title _____

Institution/Department_____

Address _____

E-mail Address_____

Librarians, please send your orders directly to LEA or contact from your subscription agent.

Lawrence Erlbaum Associates, Inc., Journal Subscription Department
10 Industrial Avenue, Mahwah, NJ 07430; **(201) 258–2200;** FAX (201) 760–3735; journals@erlbaum.com

THE POWER OF WORDS
Unveiling the Speaker and Writer's Hidden Craft
David Kaufer
Suguru Ishizaki
Carnegie Mellon University
Brian Butler
University of Pittsburgh
Jeff Collins
Carnegie Mellon University

In 1888, Mark Twain reflected on the writer's special feel for words to his correspondent, George Bainton, noting that "the difference between the almost-right word and the right word is really a large matter." We recognize differences between a politician who is "willful" and one who is "willing" even though the difference does not cross word-stems or parts of speech. We recognize that being "held up" evokes different experiences depending upon whether its direct object is a meeting, a bank, or an example. Although we can notice hundreds of examples in the language where small differences in wording produce large reader effects, the authors of *The Power of Words* argue that these examples are random glimpses of a hidden systematic knowledge that governs how we, as writers or speakers, learn to shape experience for other human beings.

Over the past several years, David Kaufer and his colleagues have developed a software program for analyzing writing called DocuScope. This book illustrates the concepts and rhetorical theory behind the software analysis, examining patterns in writing and showing writers how their writing works in different categories to accomplish varying objectives. Reflecting the range and variety of audience experience that contiguous words of surface English can prime, the authors present a theory of language as an instrument of rhetorically priming audiences and a catalog of English strings to implement the theory. The project creates a comprehensive map of the speaker and writer's implicit knowledge about predisposing audience experience at the point of utterance.

The book begins with an explanation of why studying language from the standpoint of priming—not just meaning—is vital to non-question begging theories of close reading and to language education in general. The remaining chapters in Part I detail the steps taken to prepare a catalog study of English strings for their properties as priming instruments. Part II describes in detail the catalog of priming categories, including enough examples to help readers see how individual words and strings of English fit into the catalog. The final part describes how the authors have applied the catalog of English strings as priming tools to conduct textual research.

Contents: T. Oakley, Foreword. Preface. Introduction: Words and Their Potency for Priming Audiences. **Part I:** *Preliminaries.* Priming Audience and Practices of Literacy. Cataloging English Strings for Their Priming Potencies: A Report of a Research Study. Methods for Selecting and Catloging Strings. The Catalog Hierarchy. The Hierarchy in Relation to Previous Scholarship. **Part II:** *Results: The Catalog in Depth.* Cluster 1: Internal Perspectives. Cluster 2: Relational Perspectives, Part I. Cluster 2: Relational Perspectives, Part II. Cluster 3: External Perspectives. **Part III:** *Implications and Applications of Rhetorical Priming Theory.* Using Priming Strings to Analyze Corpora of Texts.
0-8058-4783-9 [cloth] / 2004 / 272pp. / $59.95
Special Discount Price! $32.50
Applies if payment accompanies order or for course adoption orders of 5 or more copies.
No further discounts apply.
Prices are subject to change without notice.

Lawrence Erlbaum Associates, Inc.
10 Industrial Ave., Mahwah, NJ 07430–2262
201–258–2200; 1–800–926–6579; fax 201–760–3735
orders@erlbaum.com; www.erlbaum.com

ATTW CODE OF ETHICS

The Association of Teachers of Technical Writing is the principal national organization of teachers of technical communication. We have unique responsibilities as academic professionals and general responsibilities as technical communication professionals.

Our code of ethics articulates these responsibilities. It states what we expect of ourselves and what others may expect of us while it offers specific points of reference for dealing with ethical situations. By nature it must be general and so requires responsible interpretation in any particular application.

Our core principles are presented first, followed by descriptions of these principles in practice. These sections address our activities as both teachers and professional technical communicators.

Core Principles
- To act honestly, fairly, responsibly, and professionally in our relationships with students, colleagues, employers, and research subjects.
- To provide clear, accurate, appropriate, and effective technical communications.
- To recognize the power of language to shape thoughts, values, and actions and to accept responsibility for the likely consequences of our language.
- To accept our responsibilities to the public for our technical communications.

Principles in Practice
The ways we put these principles into practice are grouped by relationship because ethics deals with our responsibilities in relating to others.

To Students
- To foster a sense of ethical responsibility to themselves, stakeholders, and the public.
- To prepare them for technical communication careers with appropriate theories, knowledge, and skills.
- To respect them as individuals entitled to fair, equal, and helpful interactions and as professionals themselves.

To the Public
- To promote the understanding of the role and significance of technical communication in society.
- To develop technical and scientific literacies for informed, critical decision-making.
- To make our technical communications understandable to all those reasonably affected by them.
- To protect the security, confidentiality, and privacy of the information we are entrusted with.
- To adhere to standard principles of research with human subjects by obtaining informed consent and maintaining the privacy and confidentiality of research results.

To the Technical Communication Profession
- To advance the profession through continuing research, education, and training in the technologies of our field.
- To maintain the reputation of the profession by acting responsibly in teaching, research, consulting, and writing.
- To encourage the recognition of the teaching of technical communication as a valuable and tenurable activity.
- To promote increased participation by under-represented groups in academic and nonacademic technical communication.

To the Academy
- To promote the academic traditions of advancing and sharing knowledge, tolerating diversity of opinion, offering responsible criticism, and encouraging freedom of expression.
- To promote intellectual property rights and fair use exceptions to facilitate the educational and critical use of information.

To Non-Academic Employers and Contractors
- To develop technical communications that aim to satisfy their needs, goals, and interests.
- To develop technical communications that are clear, effective, efficient, appropriate, accurate, useful, and delivered on schedule and within budget as nearly as possible.
- To respect their legitimate rights to confidentiality, security, and profitability in a competitive environment.
- To charge fairly, honestly, and appropriately according to mutually understood agreements.
- To make as clear and explicit as possible our contractual obligations, responsibilities, and expectations.
- To honor the terms of our contracts and to negotiate in good faith when disagreements arise.
- To improve technical communication practices in the workplace through continuing research, education, training, and dissemination of knowledge.

THE STATE OF TECHNICAL COMMUNICATION IN ITS ACADEMIC CONTEXT, PARTS I & II

A Special Issue Set of *Technical Communication Quarterly*
Guest Editor

Carolyn Rude
Virginia Tech

Several recent developments in the history of the Association of Teachers of Technical Writing (ATTW) and *Technical Communication Quarterly (TCQ)* led to two special issues on the state of technical communication in its academic context. These issues focus on the work of the ATTW as it helps to guide the evolution of the field, including a description of its members, reflections on the journal and its history, assessment of student learning, research in the field, and the academic job market. The articles are written by members of the Executive Committee and their collaborators and others who are leaders in a particular subject area. The ATTW Executive Committee has taken the opportunity that change offers for some self-study and reflection on the field and the role of academics in it. This reflection will help leaders of the association and academics in general to develop a vision and plan for the future.

The two issues together provide historical background on the Association of Teachers of Technical Writing (ATTW) as well as data on current academic members of the field and their jobs, their teaching and research, and their programs. The focused goal of the issues is to help ATTW plan for the future by identifying needs, interests, and responsibilities of members, and the broader goal is to define the values and current practices as well as the visions and needs of technical communication in its academic context.

Volume 13, Number 1, 2004. Contents: C. Rude, GUEST EDITOR'S COLUMN. C. Thralls, M. Zachry, INTRODUCING THE NEW TCQ EDITORS: *TCQ:* A Vision of the Future. ARTICLES: D. Dayton, S.A. Bernhardt, Results of a Survey of ATTW Members, 2003. S. Harner, STC's First Academic Salary Survey, 2003. C. Rude, K.C. Cook, The Academic Job Market in Technical Communication, 2002-2003. A. Blakeslee, R. Spilka, The State of Research in Technical Communication. J. Allen, The Impact of Student Learning Outcomes Assessment on Technical and Professional Communication Programs. M.M. Lay, Reflections on *Technical Communication Quarterly*, 1991-2003: The Manuscript Review Process. D.H. Cunningham, The Founding of ATTW and Its Journal. REVIEW: T. Bridgeford, *Reshaping Technical Communication: New Directions and Challenges for the 21st Century*, Edited by Barbara Mirel and Rachel Spilka.
0-8058-9565-5 [paper] / 2004 / 144pp. / $35.00

Volume 13, Number 2, 2004. Contents: C. Rude, GUEST EDITOR'S COLUMN. ARTICLES: S. Selber, The CCCC Outstanding Dissertation Award in Technical Communication: A Retrospective Analysis. N. Allen, S.T. Benninghoff, TPS Program Snapshots: Developing Curricula and Addressing Challenges. L. Gurak, A.H. Duin, The Impact of the Internet and Digital Technologies on Teaching and Research in Technical Communication. R. Johnson-Sheehan, C. Paine, Changing the Center of Gravity: Collaborative Writing Program Administration in Large Universities. K.T. Rainey, R.K. Turner, Certification in Technical Communication. REVIEWS: M.S. Salvo, *Power and Legitimacy in Technical Communication Volume I: The Historical and Contemporary Struggle for Professional Status*, Edited by Teresa Kynell-Hunt and Gerald J. Savage. L.F. Gattis, *Openness, Secrecy, Authorship: Technical Arts and the Culture of Knowledge From Antiquity to the Renaissance*, by Pamela O. Long. A. Ilyasova, C.E. Ball, *Writing Space: Computers, Hypertext, and the Remediation of Print*, 2nd ed., by Jay David Bolter.
0-8058-9552-3 [paper] / 2004 / 111pp. / $25.00

Lawrence Erlbaum Associates, Inc.
10 Industrial Ave., Mahwah, NJ 07430–2262
201–258–2200; 1–800–926–6579; fax 201–760–3735
orders@erlbaum.com; www.erlbaum.com

CALL FOR PROPOSALS

SPECIAL ISSUES OF
TECHNICAL COMMUNICATION QUARTERLY

The Association of Teachers of Technical Writing (ATTW) is seeking proposals for special issues of its journal, *Technical Communication Quarterly (TCQ)*.

Special topics could include, but are not limited to, **communication design, the role of digital technologies in technical communication, the rhetoric of workplaces or professions, pedagogical approaches, the practices of publication management, dialogue between academics and practitioners, research methods in the field, and connections between social practices and organizational discourse.** A list of previous special issues and their editors is available at **www.attw.org.**

Proposals will be reviewed by the ATTW Executive Committee, and the proposing editor(s) will be contacted. Please submit proposals by postal mail or e-mail to one of the journal's special issue editors, Sherry Burgus Little or Richard Johnson-Sheehan.

Sherry Burgus Little
Department of English
San Diego State University
San Diego, CA 92182-8140
U.S.A
(619) 594-5238 / Office
(619) 594-4998 / Fax
slittle@mail.sdsu.edu

Richard Johnson-Sheehan
Department of English
University of New Mexico
Albuquerque, NM 87131
U.S.A.
(505) 277-4144 / Office
(505) 277-5573 / Fax
rsheehan@unm.edu

Printed in the United States
by Baker & Taylor Publisher Services